T0305634

The Secrets to Construction Business Success

With a daunting industry-wide business failure rate, construction professionals need to manage risk and finances as effectively as they manage projects and people. *The Secrets to Construction Business Success* empowers contractors and other professionals to defy the long odds threatening their stability, growth, and very survival. Drawing on the authors' more than eight decades of combined experience turning around failing firms, this book provides a master class in structuring, managing, and futureproofing a construction business. Chapters on measuring and responding to dips in revenue equip executives to recognize and respond to the warning signs of financial distress while chapters on succession planning ensure that organizations survive their founders' departures. Sample documents and tools developed for the authors' consulting practice offer field-tested solutions to organizational structure, forecasting, and accounting challenges.

A steady source of guidance in an industry with few constants, *The Secrets to Construction Business Success* makes an invaluable addition to any industry leader's library.

Thomas C. Schleifer, PhD, is a turnaround expert and the founder of the largest international consultancy firm serving the contract surety industry. He was also an Eminent Scholar at Arizona State University's Del E. Webb School of Construction. He joined the construction industry at age 16 and has amassed more than 60 years of experience. This combination of practical, hands-on experience as a contractor and assisting financially distressed companies has given Dr. Schleifer a unique perspective on the causes of business failure and how to avoid them.

Mounir El Asmar, PhD, is Associate Professor at Arizona State University's Del E. Webb School of Construction and Co-Director of the National Center of Excellence on SMART Innovations. His non-traditional background in civil engineering, mechanical engineering, construction, economics, and business provides him with an understanding of a variety of facets imperative to advancing today's engineering and construction industry. Dr. El Asmar's work has supported decision-making for governments and private organizations in the US, Canada, Europe, and the Middle East.

The Secrets to Construction Business Success

Thomas C. Schleifer and Mounir El Asmar

Routledge
Taylor & Francis Group

NEW YORK AND LONDON

First published 2022
by Routledge
52 Vanderbilt Avenue, New York, NY 10017

and by Routledge
2 Park Square, Milton Park, Abingdon, Oxon, OX14 4RN

Routledge is an imprint of the Taylor & Francis Group, an informa business

© 2022 Thomas C. Schleifer and Mounir El Asmar

The right of Thomas C. Schleifer and Mounir El Asmar to be identified
as authors of this work has been asserted by them in accordance with
sections 77 and 78 of the Copyright, Designs and Patents Act 1988.

All rights reserved. No part of this book may be reprinted or reproduced
or utilised in any form or by any electronic, mechanical, or other means,
now known or hereafter invented, including photocopying and recording,
or in any information storage or retrieval system, without permission in
writing from the publishers.

Trademark notice: Product or corporate names may be trademarks
or registered trademarks, and are used only for identification and
explanation without intent to infringe.

Library of Congress Cataloging-in-Publication Data
Names: Schleifer, Thomas C., author. | El Asmar, Mounir, author.
Title: The secrets to construction business success / Thomas C. Schleifer
and Mounir El Asmar.
Description: New York, NY : Routledge, 2021. | Includes bibliographical
references and index.
Identifiers: LCCN 2021034068 (print) | LCCN 2021034069 (ebook) |
Subjects: LCSH: Building—Vocational guidance. | Construction
industry—Management.
Classification: LCC TH159 .S35 2021 (print) | LCC TH159 (ebook) |
DDC 690.023—dc23
LC record available at https://lccn.loc.gov/2021034068
LC ebook record available at https://lccn.loc.gov/2021034069

ISBN: 9781032135106 (hbk)
ISBN: 9781032134734 (pbk)
ISBN: 9781003229599 (ebk)

DOI: 10.1201/9781003229599

Access the support material: www.routledge.com\9781032134734

Typeset in Goudy
by codeMantra

To Sophann and Nathalie.

Contents

Figures

Tables

Preface

The success of a construction business starts with avoiding failure. Understanding the reasons why construction businesses fail is the best way to prevent unnecessary loss. The investigation and resolution of hundreds of construction company bankruptcies has generated a fascinating understanding of the subject. The events and decisions that precede a business collapse can be categorized and quantified to define the most common causes of failure. We have been privileged with the opportunity to analyze construction business failures over five decades; we wrote this book to provide a blueprint for how to run a successful business while avoiding some of the most common landmines.

Many construction companies spend considerable time and effort on managing projects; sometimes there is not enough focus on managing the business entity itself. This book does not cover *project* management topics and construction field operations (for which abundant literature exists), but instead focuses on *business* management for construction, for which there is a major gap and a real industry need. The presented paradigms, patterns, and models were researched and developed based on insights from successful careers as a contractor, a founder of the largest surety consultancy in the country, a college professor, and a researcher developing national guidance for government and industry. This unique perspective is distinctive in our construction industry and allowed the authors to research, discover, and confirm new concepts that they share in this book.

In addition to road-mapping how to profit and prosper in the construction business, the authors also share the essence, secrets, and philosophy behind the success principles of leadership and management of a construction enterprise. The book will offer more than how these principles work, but why they work, providing not only the mechanisms but also the background and scholarship around each subject. The reader will be able to master the concepts and own the solutions by becoming able to create and develop resolutions using what they learn. The authors take the mystery out of the business side of construction. For the first time in one place, these original concepts and processes will be presented in great detail. They include flexible overhead, construction corporate self-analysis, indicators of financial distress, mitigating hidden risks, the science of project selection, subcontractor management, how to manage customer relations, getting paid on time, prospering in cyclical markets, construction industry consolidation, construction company stages of growth, succession planning, new market entry in construction, and liquidity management, among others.

The book offers a practice-oriented look at the business side of construction enterprises; numerous tools and templates are included so that the reader can implement the concepts directly in their construction business. Links to a number of proprietary interactive programs and spreadsheets developed by the authors are included in this book and offered at no additional cost, such as the Project Selection Program, Construction Company Corporate Self-Analysis Program, Contractor Financial Self-Analysis, and R-Score Calculator. Also included are actual samples of a Business Plan, Succession Plan, Subcontractor Management Policy, Change Order Management Policy, Self-Directed Professional Development Plan, Self-Administered Skills Inventory, and more.

To summarize, this book is about how to be successful, how to avoid the pitfalls that look like opportunities, and how the construction business works. Everyone involved in the business of construction will benefit from this book – from those engaged with the largest firms in the nation to those in smaller construction enterprises. Construction leaders and professionals from all construction-related professions will learn more about their industry, how to advance in their careers, and how to lead a successful construction business.

Part I

Structuring a Construction Business

1 The Basics of Construction Business Success

The construction industry has changed dramatically over the years and continues to become more sophisticated, employing new technologies and innovative processes at an astonishing rate. Construction has also moved closer to becoming a commodity, which is a primary reason for profit margins being low compared with historic norms. Lower margins increase risks, allowing less room for error. Profit enhancement in the future will depend primarily on productivity improvements and efficiency, as well as integration in design and construction to streamline the delivery process and reduce inherent waste.

Management decisions are arguably the most critical determinant of the success or failure of a construction business. However, many construction professionals believe their company loses money or fails because of other reasons such as labor problems, weather conditions, inflation, interest rates, cost of equipment, tightening or shrinking of the market, or simply bad luck. None of these alone are primary causes of contractor failure. They significantly contribute to financial distress once a poor management decision is made, but are not the basic causes of failure. Failure is often not a result of factors or conditions over which management has no control. Company leadership is responsible for structuring a company that is flexible and can withstand fluctuations in labor and market conditions, equipment and financing cost, and so on. Moreover, leadership has a critical role to make decisions that do not expose the company to such variability and risks unnecessarily.

Even if a company is successful, management cannot drop their guard. Growth or simply ongoing operations involve change, which has risks associated with it. Therefore, past success is not an indicator of future success. When a company expands in size, takes on larger projects, or goes after projects of different kinds or in different territories, these activities require good management decisions to reduce the risks inherent in such change. An enterprise may be doing fairly well, or even very well; however, the organizational stress of growth or change can cause weak components to become fatal. Change itself must be managed to control risk.

There seem to be many ways to run a construction enterprise successfully, perhaps as many ways as there are companies. In reality, there are only a few ways to structure, operate, and manage a construction business successfully while controlling business risks. Success requires appropriate structure, proper organization, direction, planning, and risk recognition.

DOI: 10.1201/9781003229599-2

At a high level, a construction enterprise has only three primary functions: getting the work, doing the work, and accounting for the work. In other words, the key functions are (1) estimating and sales, (2) construction operations, and (3) administration including finance and accounting. These three functions are separate and distinct but equal in importance. To be managed appropriately and effectively, they should be analyzed separately, with time and energy budgeted to each function. One key individual must have direct responsibility for each of these three functions. It is not unusual for a small enterprise to have one person handling operations while another person handles accounting for the work. Neither is it unusual for one person to handle all three functions; however, the functions remain distinct. Responsibility must be recognized. Some may consider one of these functions to be more important, but neglecting any of these three functions is courting failure. They are equally essential to success. Individuals who accept responsibility for one or more of the three primary functional areas of management are key to the organization whether or not they own a piece of the company. The leading individual must believe he or she is ultimately accountable for the success of their functional area and accept responsibility personally, not as a functionary or an executive but as a "principal" in the organization – with ownership or not.

The success of a construction business starts with avoiding failure (Figure 1.1). Understanding the reasons why construction businesses fail is the best way to prevent unnecessary loss. The investigation and resolution of hundreds of construction company failures has generated a significant body of knowledge on the subject. The events and decisions that precede the failure of a construction

Figure 1.1 The success of a construction business starts with avoiding failure.

business can be categorized and quantified to define the most common root causes. The following section discusses the five primary elements commonly found when analyzing construction business failures.

The Five Primary Elements of Construction Business Failure

One of the most interesting phenomena is the fact that the events and decisions that cause or contribute to a construction business failure take place during the company's profitable years. To look for the causes within the difficult years (i.e., when a company is losing money or breaking even) is to study the result and not the cause. It is easy to be misled in a study of bad years because losing operations can generate unusual events and decisions, even if the contractor is unaware of the impending loss.

The events and decisions that precede the failure of a construction company generally take place one year to three years *before* the first financial statement that shows a break-even year or a loss. A study of the events and decisions that caused hundreds of companies difficulties, identified five recurrent and industry-wide elements contributing to risk and often resulting in failure. The primary common elements of construction business failure are:

1   Significant increase in project size
2   Unfamiliarity with a new geographic area
3   Moving into a new type of construction
4   Changes in key personnel
5   Lack of managerial maturity

Each of these key elements will be discussed using examples of how they affect an organization and its ability to make a profit.[1]

Usually, decisions to grow the business are consciously made, and the events are recognizable and may appear to be routine business occurrences. Many contractors deciding to take on a very large project, to expand into unfamiliar locations or to establish a new type of construction, may not see such decisions as dangerously risky; with proper planning and controls, they don't need to be. There is no suggestion here that a contractor should fear growth or other changes. What is expressed is that at least one and usually two or more of these events or decisions preceded the failure of a large number of contractors and that there is an inherent danger in these elements. A complete understanding of the risks involved is necessary when encountering them. When two or more of these business changes are undertaken at the same time, they are often lethal.

Element #1. Significant Increase in Project Size

The most common element among contractors who fail is a dramatic increase in the size of projects undertaken. The change to larger projects usually occurs during profitable years, and problems sometimes develop even before the first large project is completed.

Undertaking larger projects is a natural part of the growth of a construction company; the order of magnitude addressed here, however, is two times or greater the previous largest project. The size of a project relative to the size of the company, and relative to the size of its normal or average projects, has a definite and direct relationship to profit potential.

When a construction enterprise is operating at a profit, doing a certain average-sized project and a certain top size, there is no reason to believe that it will profit if it takes on dramatically larger work. A construction firm may be able to build a project that is two or three times larger than it normally does; however, the issue is whether they can build it at a profit. If a company can construct $1-million road projects or buildings, it may be able to construct a $2-million or a $3-million road project or building and get the job done. But the critical question is: will it make a profit? Making a profit on a job twice the size of a company's previous largest project would be unlikely. Making a profit from a job three times greater than the largest ever built is almost impossible without additional financing resources and a tremendous amount of careful planning, all of which is unlikely without outside help. Getting additional resources might be possible, but how would a contractor, with no background on projects of such magnitude, determine what resources would be needed? Without previous experience, how could they carefully plan the work? Contractors who normally do top-sized jobs of $1 million, $10 million, or $100 million would be working in an altogether different environment than the one they are equipped for if they took on a $3-million, $30-million, or $300-million job.

Let's consider an example from a case study conducted by the authors. A contractor's previous largest project is $1 million and usually has two or three additional jobs at any given time of $300,000 to $1 million and a number of smaller jobs under $300,000 each. The company's annual volume is $3 million, and it is generating a comfortable profit. When work dried up and backlog fell off dramatically, they went after larger and larger projects. They were able to capture a $3-million project, and in their estimation their problems were over for a while. In fact, their problems were just beginning. Let's look at the impact on their organization. Previously projects took about a year or less to complete. On average, one of their larger projects started about the time another finished and a third was at its midpoint. Normally, on this company's projects near-completion, there were considerable retainage, while the ones in the middle stages were generating large monthly payments, and the ones starting up were producing good cash flow through front-loading. By handling jobs in sizes they were accustomed to, which normally were in varying stages, the company not only had a reasonable cash flow but also had the time and resources available to look after all of its jobs and keep the jobs profitable.

Contrast this with the $3-million job. At first, the front-load was terrific, but the retainage mounted fast and, within six or eight months, became a higher amount than the company had ever had on all jobs combined. By the end of the job, the retainage amount was strangling the business, and the project took longer to finalize than anything they had ever undertaken. While the project type was similar to the organization's previous work, they were surprised at the level of inspection and supervision they were subjected to by the architects or engineers.

On larger projects, municipality, state, and lender inspections generally have more red tape than smaller jobs, which may be more than management is familiar with or more than what field staff can effectively handle. Work rules are often more comprehensive on larger jobs, and security and safety requirements broaden.

The larger project, although similar to other jobs the organization had performed, was not within its experience or capability to finance. The company got the job done, but making a profit was another story. With losses out-of-pocket and huge retainage outstanding, the company could not pay its bills and is no longer in business.

Element #2. Unfamiliarity with a New Geographic Area

Unfamiliarity with a new geographic area in which the contractor has not worked is a common element preceding construction business failure – almost as common as the change in project size. A contractor's primary area maybe one county, half a state, five states, or a whole country. Regardless of size, the primary area is the area in which the organization has regular operations, has developed comfortability, and has been profitable. While there are many good reasons for a company to expand into new geographic areas, such as normal growth, lack of work in their primary area, and special opportunities, the risks must be recognized and planned for. Again, the question is not whether the organization can build a similar product in a different location. Rather, it is whether this activity will result in profit.

With time, an organization becomes accustomed to working in a given area and can easily assume its type of work is performed the same way everywhere. Yet the differences in customs, methods, procedures, regulations, work rules, and labor conditions can be significantly different and expensive if not planned for. Examples are numerous. Consider a merit shop contractor bidding outside its area without knowing in advance that the work would have to be performed using union labor. Or, in certain areas of the country, it is common to install underground pipe practically underwater while specifications in other areas require complete de-watering. In some states, it is almost impossible to keep full crews during the first week of the deer-hunting season. There are even some areas where local suppliers will give their best prices and services only to local contractors. Regulatory requirements and inspections may differ greatly between a city center and the suburbs, or in a neighboring county.

Without even going into geological and weather conditions, there are enough potential differences to cause a prudent contractor to ensure they know what they are getting into when taking on work outside their customary area. Local help, such as a joint-venture partner or new area-based personnel, may be needed to facilitate the project. Compounding the problem, a contractor often takes a distant project, much larger than anything they have built in the past because it wouldn't pay to take projects of their normal size so far away, which of course magnifies the risk.

Element #3. *Moving into a New Type of Construction*

Contractors sometimes change from one type of construction to another, or add a new type of work to their existing specialty. Companies may change, for example, from highway work to sewage treatment plants, from heavy industrial to tunnel work, from low-rise to high-rise buildings, or from office buildings to hospitals. The need for research and planning before embarking on a new type of construction work is well recognized by contractors. What is often underestimated is the entrance cost, i.e. the costs associated with the learning period during which an organization adjusts to performing a new type of construction work. Hiring a person who masters the new type of work may not be enough. Companies often complete one or more losing jobs before they can execute a new type of construction profitably. Unfortunately, some companies do not survive this period.

Contractors are more specialized than they realize. Some construct several types of projects, for instance, but perform and profit better at one kind. Some may call it luck, but it's probably because they are better at pricing and constructing that type of work. Contracting organizations usually start and remain with types of construction in which they have the expertise. Further, they will base growth and success on the continued perfection of that expertise. Over time they become better able to estimate their kind of work and, therefore, become more competitive at getting it. They also become better at organizing and putting the work in place and become more profitable at doing it. Being able to plan and execute the construction of a bridge does not mean the same team can profitably plan and execute a building.

A more subtle change in the type of work is the change from public to private sectors, or vice versa. This change, even when the project is a company's normal size and in their area, has cost numerous firms a great deal of money. It certainly can be done with a healthy respect for the differences and risks involved and good planning, but the odds are worse if the firm has never done it before. Indeed, many companies do both public and private work and have been doing so profitably for years. There is no suggestion here that it shouldn't be done, just a report that many contractors did not recognize the differences in advance and proceeded to price and produce the work for a loss. There are considerable differences between public and private work, such as qualifying for bid lists, differences in criteria used for selecting the contractor, and expectations about change orders. The differences in expectations on public and private work may not even be known by the parties, possibly resulting in disputes that could have been avoided.

Qualifying for bid lists works differently in the two sectors. In public work, bidders may need to pre-qualify with the public body, the state, or other agencies, but the lists are often open to all contractors. A lot of startup contractors achieve their growth within the public sector. Their size of a project may be restricted at first by bonding requirements, but once they have pre-qualified, they have a good source of work. This is one of the reasons public jobs usually have more bidders than private jobs.

Most private sector work, on the other hand, involves select lists that are more difficult to get, as owners or designers may pick the preferred contractor, sometimes in a less transparent manner because the project is privately funded and isn't bound by the same requirements that govern when taxpayer money is involved. Few startup contractors can find their way onto the private sector select lists where the number of proposers is usually fewer than on public projects of similar size. The number of competing contractors on a project statistically affects the number of projects a company has to go after in order to be awarded one. This impacts the cost of doing business, which affects profit margins.

While regulations are changing with the rise of alternative project delivery methods, which allow qualifications to drive contractor selection, many public organizations are still required by regulation to award a significant portion of the work to the lowest bidder. This traditional low-bid type of contractor selection allows only limited control over who gets the work. The parties are often strangers; therefore, the awarding and administration of projects are at a significant distance from one another. A public project is administered "by the book," meaning the contractor intends to perform work according to the specifications. Because future contractors are selected based on the lowest bid for public work, there are no significant incentives for current contractors to perform beyond a project's set expectations.

The opposite is true of private work where the awarding party picks the proposers, may or may not open proposals publicly, and may often end up working with a known or at least a pre-selected contractor. The owner, architect, and contractor are much more likely to collaborate on a private project. The contractor is incentivized to perform, innovate, and deliver above and beyond the requirements because this can impact them being selected for future jobs. The same can be said in alternative delivery method projects, such as design-build (DB) and construction manager at risk (CMAR), if the selection of the contractor relies heavily on qualifications and prior experience in addition to the price. The incentives change and drive different behaviors.

Element #4. *Changes in Key Personnel*

As mentioned earlier, there are three primary functional areas of a construction business, and each must be adequately managed and supervised in a successful contracting enterprise. The primary functional areas are estimating and sales (getting the work), construction operations (doing the work), and administration and accounting (accounting for the work). In every successful construction enterprise, a top-level executive is responsible for each of these areas. Depending on the size of the company, one person could be responsible for all of them or two people may share the responsibilities (e.g., one person leads operations and the second person leads sales and administration). If a company is making a profit, it is to a large degree because of the efforts of these key individuals. If one of them leaves, there is by definition no track record of profitability for the new organization as it is reconfigured. This is a simple reality in business and even more so in the construction business that is so often defined by closely held small or medium-sized companies.

Some may point to a business with six or eight good project managers and say, "that's why this company makes money." But someone may also point to the person who is primarily responsible for construction operations and say, "this company has those six or eight good project managers because of him or her." The same can be said about two or three key estimators, and some will say the same about the person primarily responsible for getting the work. Successful companies relegate responsibility for the primary functional areas of their companies to key people. Next, the authors briefly discuss each of the three primary functional areas, starting with construction operations (Figure 1.2).

The loss of a profit-making top manager puts the company at risk. The top management team of a construction enterprise is small compared to other industries because the labor side of the business is field-managed, and some contractors even subcontract a majority of the fieldwork. The corporate organization is separate and distinct from the field organization. The quality of field management often relies primarily on the quality of the key person or persons responsible for construction operations. Operations are the core function and provide the entire cash flow for the company. If a key person in charge of construction operations leaves the organization, the company is permanently changed and at risk until his or her replacement proves they can do the work for a profit.

On the estimating and sales side of a construction business, one or more key individuals will be responsible for the firm's pricing strategy. This manager(s) will usually take a firsthand part in bid preparations and will determine the final price. The takeoff and estimating staff may be a great asset to the company, but the top manager(s) put them together and are ultimately responsible for the success or failure of capturing the work. If this key manager(s) leaves the company, the organization no longer has a proven team that can get the work, until shown otherwise.

The areas of administration and accounting are often overlooked and underrated by contractors. If there are only two top individuals in the organization who are responsible for the three primary functional areas of the business, one of

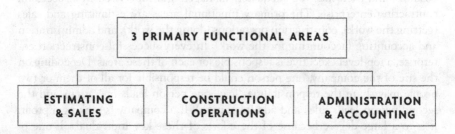

Figure 1.2 The three primary functional areas of a construction business.

them will be responsible for the administration and accounting functions. Usually, these fall to the person responsible for getting the work because sales and estimating are more of an office function than construction operation.

In smaller organizations, it is sometimes difficult to determine who is in charge of administration and accounting because this function is often not recognized as a primary area critical to a company's success. It is often relegated to middle managers even in medium and large-sized companies, which is less than ideal. This problem is most acute in growing, medium-sized firms. When the business is small, the contractor may run the entire business, including such details as signing checks. Therefore, the contractors are close to the accounting side, if only to pay the bills and be informed of the bank balance. If borrowing is required, they are the ones who explain it to the banker. Administrative needs are few, compared to a larger contractor. The small contractor may or may not keep minutes of important meetings, confirm agreements in writing, or even reply to all correspondence received. The small contractor is in continuous communication with the (relatively few) players on their work in progress, and as a result, the impact on the business of poor paperwork and administration is reduced. As the company grows and the staff increases, administrative and accounting duties often continue to be relegated to middle managers.

If a principal in a construction firm is not responsible for this important primary function, the enterprise may be improperly managed. If a dedicated, capable manager who takes personal responsibility for the administrative and accounting functions cannot be identified, the company may have a serious problem and is not organized for success.

Along the same lines, if the person who was ultimately responsible for the company's administration and accounting functions during profitable years transitions out of the organization, then the organization is at risk. The accounting staff, under new management, has no track record for monitoring the company's progress and developing accurate and meaningful financial information.

In summary, one common cause of company failure is an inadequate replacement of the person or persons responsible for one of the three primary functional areas of the construction enterprise. Typically, the changes in key personnel that contribute to, or causes problems, take place while the business is profitable.

Element #5. Lack of Managerial Maturity

This element of contractor failure is often found in conjunction with one or more of the other elements of business failure and may be a contributing cause of the other four elements discussed so far. Many construction organizations were founded by one person. The entrepreneurs who survive the high mortality rate for startups usually enter a growth stage. The qualities and abilities required for a contractor to succeed at a small construction business are not necessarily the same as those required for the success of a larger construction business. Confidence and independence, the very traits that cause an entrepreneur to want to be in their own business, to begin with, can mask the risks of growth.

Many entrepreneurs assume, "If I succeeded at this volume, I'll do twice or three times as well at two or three times this volume." At some point in the growth of every enterprise, however, the organization must change; it must become more sophisticated. At these junctures more authority must be delegated, more complex systems and procedures will be required, and more sophisticated people may be needed to handle them. Most entrepreneurs seem to instinctively follow a "command and control" approach; however, as an organization expands it requires delegation and some release in control. Some entrepreneurs have great difficulty with that. It would be easier if these changes evolved slowly over a growth period because they would be easier for the contractor to digest. But this is not usually how it works because you can't hire half of a person or put in half of a new system. So, when these changes occur, they are often significant and relatively fast.

Knowing when and how to make organizational changes is an aspect of running the business that tests the true skills of the contractor in a growing firm (Figure 1.3). Organizational changes necessitated by growth need to be made during successful times to assure continued success. One key to success in management is not to eliminate all problems, but to focus on the critical problems of the present stage of the organization's life cycle, so it can grow and mature to deal with the problems of the next stage.

The contractor who resists change until they have proof of the need for change, perhaps until they have a losing year, may have waited too long. Some of the organizational changes that may be required to expand are difficult to recognize and hard for some contractors to accept.

Figure 1.3 Knowing when and how to make organizational changes is an aspect of running the business.

Delegating responsibility and authority; hiring outside top managers who may have to supervise long-time associates, friends, or family members; and sharing financial information with more people are a few of the difficult decisions a growing company faces. Even harder may be accepting concepts like open-book management where each employee learns to understand the company's financial information, along with all other numbers that are critical to tracking the company's performance.

The term "managerial maturity" is used here to mean a contractor's managerial abilities must mature as the business does. They must change from doing everything themselves to building an organization that can do everything as well, or even better than they did. Contractors who are unable or unwilling to change their organizations to deal with their growth should either curtail their growth and level-off, or face the risk of the business outgrowing its organization. Attempting to perform $100-million worth of construction work with a $20-million organization is extremely risky.

Additional Concerns and Associated Opportunities

In addition to the five primary elements of construction business failure, there are several miscellaneous areas of concern that have caused performance deterioration, or worse, for a number of construction companies. Any of these concerns, unaddressed, can induce financial distress. If an organization is suffering from any of the primary elements of contractor failure discussed earlier, these miscellaneous areas of concern amplify the problem. The degree to which an organization experiences each concern influences the impact on the company. Some of these issues may appear minor if an organization is not suffering from them; however, anyone of them can impair a company if they get out of hand. Some are difficult to discern and can exist for many years unrecognized. For that reason, these bear careful reading periodically by those responsible for the success and well-being of a construction organization. The authors identify ten areas of concern and discuss them in the following sections. The first four concerns are organization-related, the next three are growth and market-related, and the last three are people-related.

1 Business planning
2 Flexible overhead
3 Debt
4 Claims
5 Growth rate and risk
6 Market-driven, not volume-driven
7 Peaks and valleys
8 Internal company disputes
9 Employee benefits and compensation
10 Motivation and loyalty

1 **Business Planning**

Formal long-range and strategic planning is not high on the agenda of many contractors. That is not to say contractors can't express their objectives and the plans they have to achieve those objectives, but many don't have a written guide – a detailed long-term business plan. Without a detailed plan, you are forced to react on the spot to whatever comes your way instead of setting a direction, controlling the business, and measuring progress.

Short- and long-range formal written plans are the tracks on which a company runs; they make managing a construction business much easier. Developing and following strategic and long-range plans is a proven and effective tool for success. The time spent in the planning comes back to the organization with incredible interest and dividends of time saved.

Planning should be done at a time set aside for just that purpose and outside the mainstream of daily activity. The owners and key managers of the firm should discuss and evaluate their individual and corporate goals and ensure that these goals align. Perhaps not everyone envisions moving in the same direction, but everyone needs to get on the same page. All of the company's resources should be realistically evaluated and measured against short- and long-range goals to see if they fit. By establishing clear goals and directions that are understood by everyone concerned, meeting these objectives becomes easier as everyone is thinking along the same lines and looking in the same direction.

Things certainly don't always go as planned, but much of a contractor's future business is within their control. When things change, plans can be reevaluated and altered so that the organization is not reacting but acting in a structured fashion. The plan provides a measure of movement. The importance of formally written, short- and long-range, detailed plans cannot be overstated. A construction company needs a one-year, detailed business plan combined with two more years of soft or flexible plans. A shorter plan may be sufficient for a smaller or newer company. Long-term business planning is the ultimate risk control tool for contractors. Those who make the effort find that they can manage their business with fewer surprises, more confidence, and a clearly defined purpose. The concept of business planning will be discussed in greater detail in Chapter 3.

2 **Flexible Overhead**

In addition to business planning, another important area of concern that has caused companies' performance deterioration is overhead. Overhead can be generally referred to as the cost associated with running a business. Overheard can be typically categorized into two groups: indirect overhead costs which are not specific to any particular job, and direct overhead costs which are project-specific.

Flexible overhead, a new concept for the construction industry, advises construction companies to have a certain portion of their indirect and direct overhead costs easily adjustable. The marketplace is so unpredictable and affected by so many variables that it is difficult to accurately forecast for even a few years. If a growing enterprise cannot be sure of sustained growth well into the future but needs to add overhead to deal with current growth, it can control risk by contributing additional overhead that can easily be removed if the market turns down. With some of their overhead being flexible, the company does not become dependent on their volume and can fall back if necessary and concentrate on profit.

The flexible overhead method is used temporarily for employee services for some clerical, administrative, and accounting functions. Use short-term rentals for some office and field equipment and short-term office leases, even temporary trailers, during growth stages until a new plateau of volume can be reasonably ensured. Even management people can be brought on with specific company growth and performance goals associated with their continued employment. This creates healthy challenges for new team members and refocuses the real risks associated with growth for existing management. This practice has been successful with established companies, as well as startup firms, and is being embraced by a growing number of construction enterprises.

There may be minor costs associated with a flexible overhead as lease and rentals may cost more than purchased equipment, temporary employees may cost more than permanent full-time employees, and efficiency could suffer if not managed well. But the reduction and control of risk are well worth a modest additional expense. An added benefit is the motivation of existing management people who get involved and excited about this prudent, realistic, and businesslike approach to growth. Existing managers can easily discern the positive impact flexible overhead has on their job security.

Flexible overhead may create cramped quarters and less comfort than private plush offices, but those who use it to control risk during incremental growth phases say they sleep a lot better when they get home at night. Putting on permanent overhead in a fickle market is just too dangerous. Most who have tried the flexible overhead approach have been impressed with it to the degree that they are now reluctant to add permanent overhead, and when needed they do so even slower than would normally be considered safe. They are committed to keeping some portion of their overhead flexible at all times as a hedge against a market slump, and that portion seems to grow as they realize how easy and economical it is. The modest added cost, if any, is not unlike an insurance premium for protection from a known and measurable exposure. Companies that embrace flexible overhead manage their profit and not their volume.

Flexible overhead prepares a construction enterprise to do 15 to 25% less volume at any given time, while at the same time preparing it to do 25% or more work and have no permanent increase in overhead risk either way. An organization skilled in flexible overhead can gear resources up and down temporarily and more quickly and economically than an average construction company can secure permanent resources. This practice is a departure from the accepted norm, but it is clearly the profile of the successful contractor of the future. The concept of flexible overhead will be discussed in greater detail in Chapter 4.

3 Debt

An important area of concern is debt. Construction companies use credit in many ways: secured loans to purchase equipment, lines of credit to fund working capital as needed or to fund growth, surety credit to secure payment and performance bonds and so on. Arranging and managing credit is not a singular event; it is a process requiring skills and attention. Borrowing is a planned event. Unplanned or unscheduled borrowing is often a warning sign. It is too common for a company to borrow for working capital unexpectedly and not fully understand why the need arose. Management should be aggressive in determining why the money is needed today when there was no anticipation of the potential need the previous month. A construction business operating without cash flow planning is out of control for the simple reason that they never know when they are going to run out of money.

A large line of credit is no substitute for cash flow planning. A company with no debt still requires cash flow planning. Cash flow planning is even more critical for a company borrowing some or all of its working capital. Not only is there the possibility of running out of both cash and credit, but there are also interest costs to be considered.

Cash flow planning must be included in all key decisions. Primary considerations in every business decision are profit and cash flow, and some managers believe that cash flow can be more important than profit. These related questions need answers for every business decision: will this venture or project create a cash outlay or influx, how soon, and at what risk?

Credit and borrowing are important to the security of the business and intricate to controlling risk. Borrowing should be controlled by top management through careful planning that takes into account the amounts and timing of cash needs and sources of payback. Unplanned borrowing is cause for great concern because borrowing working capital unexpectedly signals that either the cash flow plan is wrong (raising the need for better cash flow planning), profitability is falling off, or there is another problem to be identified. In any case, new planning is required, and new planning on short notice should be undertaken with the same diligence as the original effort.

4 Claims

Years ago, construction professionals understood their areas of authority and responsibility without the need for arbitrators or judges. This has been all but lost to the industry. Contract documents get bigger, claims seminars get larger, and an entirely new group of services is offered to our industry – construction attorneys, claims consultants, and dispute resolution specialists. In today's contracting environment, with all parties in the construction process trying to relieve themselves of any or all liability, construction professionals are left with too few clearly defined roles. Claims consciousness can result in defensive activities, paperwork, and the expending of energy to the extent that there is less time devoted to running the work and making a profit.

We need a truce in the construction industry in which owners, designers, and contractors agree to their specific responsibility and liabilities instead of trying to avoid or pass them along to others. It probably won't happen often, but if the party who makes the mistake fixes it, there is no dispute. There will still be a cost to fix the problem but without the cost of arguing about it. Contractors must recognize the inherent risk of disputes and develop methods in advance to avoid them. The best approach includes speed and compromises instead of expanding minor disputes, and direct discussions with the parties affected in place of intervention to ensure problems do not escalate. The potential for disputes can increase with changes in project size and when working in unfamiliar areas or with unfamiliar owners and designers. Adding this exposure to the other concerns discussed in this book strongly suggests that business expansion and growth be looked at carefully and planned prudently.

When disputes do arise, they should be addressed quickly if possible. If the fault lies within the organization, it is usually cheaper to fix the problem internally rather than creating conflicts with project partners. When not at fault, ensure that the solution doesn't cost more than the problem. Try dealing directly and fairly with the parties involved before expanding the dispute. If forced to litigate or arbitrate, try to limit the dispute to the original issues and claim only real costs. The ridiculous but popular theory of throwing in everything but the kitchen sink clouds the issues that complicate the process and increase the cost of resolution. The idea of doubling everything because "they'll only cut it in half anyway" has backfired on a lot of people. What tends to occur is the doubling of the cost of resolution because it takes twice as long to weed out the excess and get back to the real numbers. Too often the exaggerated amount is cut in half, supporting and perpetuating the original theory.

A more serious consequence of protracted disputes is that the process distracts important people from their work and negatively impacts morale. There is also the outside chance that an off-the-wall verdict could break the company.

5 Growth Rate and Risk

In the volume-driven industry of construction that thrives on growth, there are failures even among the older and well-established firms. The words "growth" and "growing" recur in the study of the management of risk in the construction business because the business risks in construction are greater during growth phases. A construction company must be managed well to be successful, and in the best of times, there is a risk. A rapidly expanding construction company magnifies its risks even if it is closely and intensely managed. There is nothing wrong with building a bigger business. That is part of the American dream. But the increase in risk in the construction industry from growth alone cannot be understated and should not be overlooked.

Measuring the performance of a construction company is complicated by differences in sales from year to year and requires careful attention to the impact that volume fluctuations have on financial performance. If a company's market is not growing, growth is more difficult, but in a reasonable market, construction companies are almost always growing at some rate. The authors' research indicates that growth for a construction enterprise at a rate of more than 15% in one year should be considered substantial. Growth can be categorized as either sustained growth or incremental growth. Sustained growth over more than a couple of years compounds quickly. At 15% a company doubles in five years and triples in eight years; at 25% it doubles in three years and triples in five years; at 50% it doubles in less than two years and is five times larger in four years.

Growth requires more resources in the way of people, systems, and money. Success is measured in an organization's ability to find the necessary qualified people, have appropriate systems in place in advance of expansion, and finance the growth. The rate of growth impacts the likelihood that an organization will be able to bring qualified resources to bear on the new work in a timely fashion. The alternative is to expect existing resources to do more, but few construction organizations are known for having underutilized resources or bench strength.

As volume increases, the expanded company is untested as an organizational unit. The only reasonable test is for the new organization to operate profitably and smoothly for a minimum of a year. Sustained growth creates a situation where, if the test proves unsatisfactory, new growth has already been added during the test year and the company is looking at a second unsatisfactory year before they can roll back to their proven size and proven team. For many, at this stage, it is too late to retreat and recover.

Incremental growth instead of sustained growth may seem unnecessary and even unnatural, but it is the best way to control the inherent risk in growth beyond 15% or so. With a series of grow then test, grow then test again, a company can reevaluate and recover after a failed test instead of constant growth until they encounter a bad year from which they may or may not be able to recover. This is simply prudent risk control. With sustained growth, a company grows beyond its people and systems so often that it does not really have the same organization long enough to truly test it, and is at constant risk with an ever-changing team.

6 Market-Driven, not Volume-Driven

The ideal construction company would be organized to be market-driven, and not volume-driven. It would strive for carefully planned growth but be prepared to level-off or cut back on volume if the marketplace tightens or shrinks. It would use its markup flexibly as a competitive tool but never take break-even work just to maintain volume. In a tightening market (greater competition for the same work) or a shrinking market (less work available), the ideal construction company would price more competitively than it would in a better market and at the same time concentrate on making more profit on less work. It would have some "flexible overhead" built into the organization that could be cut immediately and would not hesitate to cut permanent overhead when necessary.

The ideal construction company is willing to reduce its overhead costs and get smaller to survive when necessary. The down cycle passes, and they will be ready for the upswing, but only if they come through intact. The large failure rate in this industry is driven in part by construction enterprises pushing at full speed ahead during weak or down markets with desperation pricing in an attempt to capture work that their competition needs as badly as they do.

During down cycles and when the company is not growing, it may be expected for the contractor to have difficulty in managing overhead costs. However, even in a rising market and growing organizations, dangers regarding managing overhead costs exist. An organization may be forced to put on overhead costs during growth in larger amounts than perhaps they would like. This can cause losses until the company grows into the overhead. Herein lies the double problem: reducing profits or losing money for a length of time because of a sudden increase in overhead to accommodate growth is dangerous, while needing additional volume as an absolute necessity to cover the increased overhead puts the company in double jeopardy.

As an organization attempts to increase market share, the price may suffer, as it is often necessary to make at least temporary price concessions to take the market share away from competitors, unless there is an exceptional boom market. Boom markets attract the attention of out-of-area companies who move in and establish price concessions to take the work away from local contractors and gain a foothold in the local market. While construction companies may not always make a conscious decision to lower their price when they need the added volume or new work, that is exactly what occurs. And when price suffers it is usually for a majority of new work, not just a small part of it, so the company ends up needing even more volume than originally planned because margins are suffering. This can easily lead to a downhill profit spiral during rapid growth, and it often does, because as an organization gets stretched there is little time for anyone to see the problem coming.

The additional growth then requires more overhead, creating temporary losses and the immediate need for even more volume. This spiral has caused numerous construction businesses to fail. If performance or profit starts to deteriorate during growth, it is often discovered after additional volume and people are taken on, and corrective measures are more difficult with everyone already stretched out. Overworked managers will be coping with the largest volume the company has handled, and some companies don't recover from this scenario. Some contractors have pursued sustained growth with no measurement of performance right up to failure. Regular measurement and close monitoring of overhead, profit, and volume are necessary to provide accurate data and inform the company's ability to be market-driven, or its management's very intentional decision to grow based on understanding the price it is willing to pay for growth.

7   **Peaks and Valleys**

There were always peaks and valleys in the construction marketplace, and when things got bad in a contractor's normal work area, they had to stick it out and do the best they could. Not that long ago, construction companies generally worked a lot closer to home because their businesses, employees, and equipment were not as mobile as they are today. Short-term rental or leasing wasn't as prevalent, and travel and relocation were more difficult. When the market was good, construction companies and all their competitors had a seller's market. Because contractors were not that mobile, they didn't go into new areas in great numbers and impact the market, so there was greater opportunity for substantial profits during good times.

The expression "They took the good with the bad" is appropriate here. The good years allowed for great earnings, and in a more conservative era, some of these extra earnings would be put away as reserves against lean years. Reserved or not, when a seller's market developed, contractors were able to generate substantially greater profits than they could under normal market conditions, which is not the case today because of increased competition resulting from greater mobility. Today, when a good market develops anywhere in the country, out-of-area contractors compete for a portion of that market, preventing the development of a seller's market and driving down prices. Very mobile nationwide contractors can follow good markets as do contractors from any area where there isn't enough work.

The net effect of greater construction industry mobility is that profit peaks are taken out of the various markets, while profit valleys remain. The opportunity for really big years is substantially reduced, and the average profit in the industry has diminished over time and shows every sign of staying down. Ease of mobility nationwide and internationally will continue to maintain competitive pressure, which, in turn, keeps prices down.

What this means to the average contractor is that without the prospect of the peak years that our grandfathers enjoyed, there is less opportunity to make up for bad years so they must control their valleys. With typically limited cash reserves, contractors can ill-afford to increase risks without controls and must manage their businesses cautiously, if not defensively. Limited profit margins require growth with prudence, testing as you go, and being prepared to withdraw from bad decisions quickly. Market cycles will be discussed in greater detail in Chapter 13.

8   **Internal Company Disputes**

The majority of construction firms are closely held companies or family businesses; therefore, internal disputes are not uncommon and can create discomfort and disruption. In a high-risk, low-margin industry where businesses often operate at high levels of intensity and energy, some conflicts can be expected. Personal problems can impact performance and profits significantly. Some contractors expect more from family members and are more tolerant of non-relatives. The best defense against conflict is open and honest communication of all parties at all levels within the organization. If continuing disharmony affects performance after management has attempted to resolve it, professional intervention is essential. Unresolved friction can fester and degenerate beyond repair and has caused untold personal anguish and distress. It can distort a successful organization and render an underperforming organization not worth saving. These issues can affect the succession of leadership even after a great deal of planning has gone into it. Leadership succession planning will be discussed in greater detail in Chapter 16.

9 Employee Benefits and Compensation

The subject of employee benefits and management perks fits well after having considered the topics of "flexible overhead" and "peaks and valleys" a little earlier in this chapter. The general and administrative costs of doing business are necessary to the running of a construction company, as are the costs of concrete and steel. Controlling these costs is imperative. The overhead cost of benefits and perks must be treated cautiously, and the best way is to be prudent in preparation for bad years, particularly during good years.

The discussion of bonuses is an intricate part of the management of overhead costs. Performance bonuses are common in the construction industry; however, many firms mismanage them (Figure 1.4). For effective performance bonuses, they must be part of a carefully considered formal compensation plan, which is fully understood by all participants. Random, unorganized bonuses, common in the construction industry, can cause more problems than they are worth. They add overhead costs spontaneously and haphazardly, and the benefits from them diminish rapidly over time. Some companies have even fallen into the trap of giving bonuses each year regardless of company performance. Bonuses tend to become regarded by employees as part of their

Figure 1.4 Performance bonuses are common in the construction industry.

wages and that they are entitled to them. Bonuses must be tied to the individual employee's performance, the profitability of the job, and the success of the entire company.

The cost of bonuses or unrealistic compensation packages, established during good years, has accelerated the decline of many companies when lean years hit. Luxury automobiles, club memberships, and pleasure trips are near and dear to anyone's heart and commonplace perks for hard-working managers in many construction enterprises. The biggest problem with these overhead expenses is that the costs to maintain them keeps going up, while loyalty and motivation may go down because the bonuses become expected and are taken for granted.

A company car is a valuable perk; often it is given in place of a $2,000 or $3,000 raise in a particular year even though it's worth much more than that. The problem is that two or three years later they are taken for granted by the valued employee who now may believe that they are underpaid by $2,000 or $3,000 compared with somebody else. Giving such perks is hard to avoid because so many organizations are doing it, but there is more value in having the highest paid people around with no perks than the lowest-paid people with great perks. A construction organization with high pay and few perks keeps their employees longer and has no trouble getting new ones to quit their lower-paid jobs to come to work for the company, because in most cases the lower paid employees have lost sight of the real value of their perks. Carefully managing perks versus wages is cost-effective in the long run and more professional.

Another problem with perks is they are often selected by the contractor as something they value, already have, or would like to have. Perks are usually given to employees without offering an alternative for different benefits or wages. Common courtesy demands gratitude so the giver has no real test of the level of appreciation or the value the employee places on the perk. For example, a conservative employee who is provided a company car, which is replaced every two to four years, may very well prefer to drive an older more economical vehicle and have the cost of the company car added to his or her salary.

Employee perks provided by well-meaning construction companies are universally undervalued by employees who, when they learn of the cost to the company, almost unanimously state they would prefer to revive the value in compensation.

One type of benefit provided to employees, which is critical to both the employee and the company, is an investment in their professional education and training. This will be discussed in greater detail in Chapter 17.

10   Motivation and Loyalty

Many closely held companies, particularly smaller and mid-size organizations, operate in a family or club-like atmosphere that many contractors believe generates loyalty and longevity. There is no credible research to suggest differently, but the practice can be expensive and inefficient, and which in the current working environment is becoming more difficult to maintain. People are changing jobs with greater frequency than ever before, and job security and loyalty aren't necessarily the top concerns of today's workforce. Multiple job experiences are more common today than remaining 15 or 20 years with the same company, as used to be the norm in the past.

Managers should look back over their company's history and recall who the key players were five or seven years ago. For many, this is an ever-changing scene and may be more so in the future. A close-knit group working in a club-like atmosphere may be comfortable, but if the players are ever-changing, a portion of the cost used toward maintaining the family atmosphere might be better spent on training replacements or reserved for recruitment. A well-managed construction enterprise is professional and businesslike with a certain amount of internal competition among managers. The contractor of the future will develop long-range plans around key positions, not just key people. Chapter 2 will delve into these key positions further and explore methods to effectively set up a construction company's overall organizational structure.

Review Questions. Check All That Apply.

1   Which of these is not a primary cause of construction business failure?
   a   Significant increase in project size
   b   Changes in key personnel
   c   Lack of the proper equipment
   d   Moving into new types of construction

2   Which of these are not primary functional areas of a construction company?
   a   Estimating and sales
   b   Understanding debt management
   c   Construction operations
   d   Labor management

3   Which of these are key differences between public and private work?
   a   Qualifying for contractor selection lists
   b   The criteria used for selecting the contractor
   c   The requirement for transparency in the contractor selection process
   d   All of the above

4   Organizational changes that may be required when expanding the business are:
   a   Easy to recognize
   b   Selling equipment
   c   Hard for some contractors to accept
   d   None of the above

5   Limited profit margins require:
   a   Growth with prudence
   b   Testing as you go
   c   Being prepared to withdraw from bad decisions quickly
   d   All of the above

Critical Thinking and Discussion Questions

1   Explain why project size matters to the potential success of a construction organization.
2   Why do the differences between public and private work matter at bid time?
3   Explain why the rate of growth of a construction company impacts risk.
4   Why are public bidding lists easier to get on, compared to private projects?
5   Explain why a construction enterprise should be market-driven, and not volume-driven.

Note

1 For a more in-depth presentation of each element, refer to the book *Managing the Profitable Construction Business*, by Thomas C Schleifer, PhD, et. al, published by Wiley, 2014.

2 Organizational Structure

Hard times teach hard lessons. For instance, the great recession of 2008 and the more recent COVID-19 pandemic of 2020 both alerted the construction industry to the fact that it cannot solely rely on booming times and unlimited growth to ensure profitability. Shrinking revenues in down cycles have inspired contractors to take a closer look at their management practices and search for ways to maintain profitability in the absence of top-line revenue growth.

During the decade preceding the recession of 2008, many construction contractors built efficient and professional organizations. When the market turned down, most were reluctant to dismantle these carefully erected organizations and began searching for top-line growth wherever they could find it. They were trying to preserve their organization for the eventual return of good times.

Unfortunately, all top-line growth is not created equal. Some projects pursued by specialty and general contractors were not profitable. They either traveled too far from their normal base of operations or took work that was outside their area of expertise. In most instances, in an effort to support their overhead organizations, they eroded profitability.

Is there a way to manage overhead in lean years without dismantling an organization? The answer is that one must learn to look at things differently and manage organizations more efficiently. A company is viewed as an arrangement of inputs or "functions" that contribute to achieving the desired ultimate corporate outcome. A function is a series of related activities intended to accomplish something such as earning revenue, customer services, work execution, management, or administration. An approach, known as functional analysis, studies the different functions of an organization.

A functional analysis creates a dynamic representation of what each position in the organization performs and how it contributes to the corporation's ultimate outcome. It traces the workflow from the initial project concept to the ultimate outcome and evaluates the efficiency and cost of each contributing function. A contractor needs to coordinate a wide variety of functions to create a single ultimate outcome. A functional analysis visually depicts the flow of these essential functions (as inputs) across the different activity areas that contribute to the outcome. The resulting chart reveals inefficiencies, conflicts, blockades, and redundancies that have a negative effect on productivity or profitability.

DOI: 10.1201/9781003229599-3

A comprehensive functional analysis approach is most suitable for medium and large-sized contractors that have several essential functions distributed across the entire organization. However, the functional analysis also provides valuable insight into organizational structuring for small-sized contractors, who can benefit from analyzing their business functions to optimize processes and minimize waste.

Construction contractors have traditionally developed their organizations on an as-need basis. In a small or midsize organization when an owner can no longer perform all the work themselves, they hire someone to help out. As the workload increases, that person hires someone to give them a hand, and so on. Companies typically grow their organization structures when an increase in work demands an increase in workers. It is natural. The authors describe this as *organic* growth.

As long as an increasing workload compels the growth of the workforce, organizational growth is typically not questioned. The workforce is always running to catch up with the volume of work, trying continually to remain efficient as a result. This kind of natural organic growth seems both logical and compelling and is rarely conceptualized after the fact.

The Problem

Sooner or later, the overhead organization tends to outgrow the workload. Department heads have hired and trained a team that they believe can handle the existing work efficiently. Often departments become the fiefdom of the department head, who then fights to preserve their organization. Research shows that at this juncture in a company's life cycle, senior management's initial response is to chase new work to cover the overhead already in place. Top-line growth as a solution to shrinking profitability has become somewhat of an automatic response. Many senior executives fall prey to this reflex. Top-heavy organizations seek new, sometimes marginally profitable or unprofitable, work to keep their employees and also prop up their top-heavy organizations. The need for ever-increasing revenues to support existing overhead can lead to poor pricing discipline, which can result in unprofitable contracts. Unprofitable contracts lead to an increased debt burden required to finance the bloated overhead, and profit is eroded further. Too often in these situations, sustaining the overhead becomes the company's priority, as opposed to profitability.

The top 25% of contractors, surveyed by a specialty contractors association in 2012, reported a 4.7% decrease in revenue per employee. However, the median contractor, who did not grow as rapidly, reported a 3.1% increase in revenue per employee. As the companies grew bigger, the revenue per employee decreased, a clear indicator that top-line growth does not necessarily lead to increased profitability. This finding highlights an exciting opportunity.

Overhead organizations initially grow in response to the demands of the workload, but once these organizations evolve into existence, executives are unwilling, or frequently unable, to "reconceptualize" the organization. Overhead organizations are not always designed specifically for the work that requires their existence. Therefore, the willingness and the ability to "reconceptualize" an organizational structure are critical steps toward ensuring the continued incremental growth of a contractor's profitability.

In most companies, the formal organizational structure is delineated by a traditional organizational chart (combined with additional verbal and written explanations). These representations usually focus on management levels, a span of control issues, and who is responsible for decision-making in each area of the business. Usually, the organizational chart also identifies the specific personnel occupying the positions. Figure 2.1 illustrates a sample organizational chart for a specialty contractor, and Figure 2.2 illustrates a sample organizational chart for a general contractor. The organizational charts presented here are examples of sizable self-performing contractors. For smaller size contractors, many of the

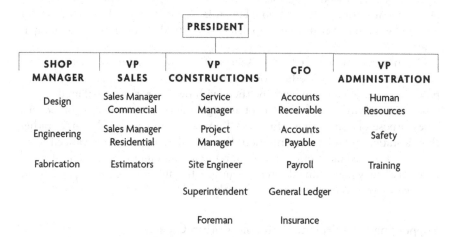

Figure 2.1 Sample specialty contractor organizational chart.

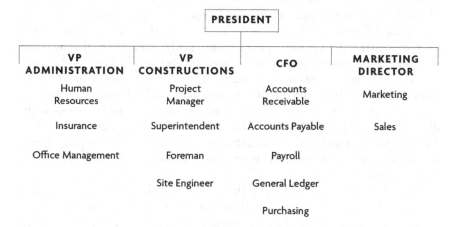

Figure 2.2 Sample general contractor organizational chart.

same functions still apply, but one person may be responsible for more than one function. In general, organizational charts can vary significantly and come in all shades and with various levels represented. There is no one-size-fits-all solution and the chart should be personalized to the company based on an analysis of its critical functions.

An organizational chart can provide information about positions and their fit in the company's hierarchy. However, it offers limited insight into the key contributions that each position makes to the ultimate corporate outcome. Traditional organization theory divides corporate organizations into departments and assumes the department title implies what it does. The inputs (i.e., functions) generating outputs and leading to the ultimate outcome are only implied but may not be directly identified or measured. Looking at the static organizational chart, no management action is suggested. It is simply a snapshot of the organization at one moment in time. The traditional organizational chart is a helpful document yet can be more powerful when supplemented with dynamic functional analysis chart.

Organization charts, like balance sheets, offer a static picture of the extent and structure of the *current* overhead organization. They are not management *"tools"*; however, management can examine them with new eyes. Functional analysis is a process identifying the parameters of an organization's functional requirements. The purpose of conducting a functional analysis is to create a dynamic, rather than a static, view of an organization. Functional analysis is an advanced technique that is used more often in larger and sometimes midsize organizations. All enterprises may have limited interest, but it is helpful to have an understanding of it even if you do not put it to use directly.

Supporting the Ultimate Outcome Within Closely held Businesses

While functional analysis is an effective tool to allow a contractor to reach their ultimate corporate outcome, a company's owner and top management play the most critical roles in leading the organization toward this outcome. The roles of a business owner and a business manager are equally significant to the success of the enterprise. However, these roles are separate and distinct because they vary based on the type of organization and whether top management owns a controlling interest in the company. For instance, in closely held companies or family businesses, owner and management roles are usually performed by the same individual, presenting a unique set of challenges. This section discusses the differences between managers and owners, their roles in supporting the organization to achieve its ultimate outcome, and practices to address unique problems that may arise in closely held businesses.

While it is possible to wear both the owner and manager hats, few business owners realize they cannot wear both hats at the same time. The roles of the owner and manager are different enough that critical decisions by management for short-term gains can often conflict with the long-range interest of the owner.

As a result, many closely held businesses suffer from a lack of clearly defined objectives and long-range strategies. Their direction tends to be defined by their management decisions first.

It is the responsibility of the business owner to establish goals and objectives for the company and to discipline the performance of the enterprise's management toward their ultimate outcome. For any sized company, a critical ingredient in the formula for success is assessing the performance of the business managers. However, the process of analyzing the efforts of management toward the achievement of predetermined goals is absent in most closely held businesses. Many owners believe that as long as everyone works diligently, puts in long hours, and cares about the success of the business, management is doing its job. Loosely translated, this means that management's "job" is to work long hours and to care, which is not always the case. Many owners consider that diligent work, long hours, and significant dedication translate to a successful business and demonstrate that management is meeting the company's ultimate outcomes. Yet, this relationship does not always occur.

Management is supposed to organize the resources of the organization to achieve attainable goals and objectives as set out by the owners or stockholders of the company. Management must also be prepared to explain, in factual detail, why they cannot or have not achieved these objectives. When the owner and the manager are the same people, these roles are in direct conflict, which hinders realistic goal-setting. In the authors' experience, this oftentimes results in the

Figure 2.3 Two hats are better than one.

company working without clear goals or accountability to meet the said goals. In this scenario, the authors like to say, "*two hats are better than one.*" (Figure 2.3)

Many owners of closely held companies refuse to have their performance measured, contending it is nobody's concern how they run their business. If asked, it is their right to respond, "that's none of your business," because it's factually true. However, they may not fully realize a fundamental definition of "running a business" includes setting objectives and measuring the performance of those charged with attaining the objectives. Not measuring performance exposes the company to risk by simply continuing to work hard until something goes wrong, or by allowing a business to grow until it self-destructs from its lack of direction.

The problem can be easily corrected. Once the differences are recognized between owner and manager, and the fact that one cannot wear both hats at the same time is accepted, the rest is relatively easier.

Owners of closely held businesses need to set aside time, a minimum of three or four times a year, to take off the "management" hat and perform the important functions of a business owner:

- Establishing and updating the short- and long-range goals and objectives of the organization,
- Reviewing and recommending policies for the enterprise, and
- Measuring the performance of management (i.e., themselves) in accomplishing clearly defined milestones and objectives.

During this activity, all parties with an ownership interest, whether active in the business or not, should be restricted to owner concerns, and not to the everyday management issues of running the business. To avoid mixing the two, these periodic sessions should be isolated from the daily business concerns by taking place away from the office. They can also take place in the office if performed after hours or with minimal interruptions because that helps facilitate objectivity and neutralize rank and title differences among owners.

When there is only one owner, the process is more difficult, if not impossible. An alternative is to develop a planning group and recruit knowledgeable, trusted individuals from within the company. The group can help establish attainable goals for the company and develop mechanisms to measure performance against these goals.

A planning group can be limited to top management only, but better results can be achieved in most organizations if middle management and the field are represented. Top managers that include labor and first-line supervisors in business planning have achieved exciting results. People who are hands-on with the work often know a lot about the strengths and weaknesses of the organization, especially when it comes to labor resources, the potential of field organization, the quality of technology, and what's right for the company.

Whether one chooses a few trusted senior managers or others from within the organization, the group should develop long-range goals for the enterprise, completely independent from their day-to-day roles within the organization.

Establishing a fundamental business planning process has a profound effect on owners and employees of closely held companies. It gets people excited about what is right for the company and how to achieve it. The concept and methods of business strategic planning are discussed in Chapter 3. Another effective method to correct the unique exposures of closely held businesses is to establish an active, independent board of directors.

Board of Advisors

A board of advisors can be an important business tool for any company (Figure 2.4). It is a group of individuals that take the owner's role as representatives of the shareholders (which may be one person or several) and have specific responsibilities. It acts as a governing body that meets at regular intervals to set policies for corporate management and oversight. A board of advisors is not synonymous with a board of directors who usually have some legal liability.

Some closely held companies have active boards of advisors or directors, but many are not effective because they are not independent of management. Boards consisting of only internal company managers, family members, or people closely associated with the business, like the company accountant, lawyer, or banker, are usually too close to everyday operations of the company. They may be biased and perhaps not objective enough to accomplish the tasks of setting attainable goals and measuring and disciplining management's performance in the pursuit of said goals. An effective board for a closely held company or family business will have several truly independent outsiders. Having at least two independent members is a starting point and can be extremely helpful to an owner who is not ready for much "outside help." However, owners who have experienced the value of independent objective outside board members often take the enlightened next step of having a majority of outside members.

Selecting outside board members is not difficult. Any person who is, or has been, successful in business is an appropriate candidate and does not need to know much about the company's specific type of business. Limited knowledge of the construction industry may be preferred, as the role of the board is not to interfere in the running of the business. They expect management to do that. The board's role is strategic concerning long-term goals, business strategies, corporate planning, and the like. It is usually not hard to find successful people in the owners' business community who would consider serving on the company's board.

Business owners can also seek nominations from managers, associates, and friends. Owners should interview candidates without obligation, to find people they trust and whom they would be comfortable working with. They should be owners or senior managers of successful businesses who had "profit and loss" responsibilities, preferably those who have been responsible for *"making payroll"* at some point in their careers. The size of the candidate's business is not critical; but if their business is significantly larger in comparison, they may not consider their appointment as important. For a small construction company, the owner of a growing printing business or successful service station may be an ideal candidate to serve on the board. Bankers with whom the company is not doing business also make ideal candidates. If the company is weak in marketing, an advertising or public relations executive would be a good addition to the board.

Outside board members are paid for their efforts, and their compensation will depend upon the size of the company. For small companies, there needs to be some compensation to establish the professional nature of the relationship. For firms under $10 million in sales, a nominal amount of one to several thousand dollars per meeting may be acceptable. For companies between $10 and $50 million, $3,000 or more per meeting might be appropriate. Some larger companies pay an annual board fee of $10,000 or more, with an additional amount per meeting.

A board should meet no less than two times per year, or it becomes too far removed from the business to be effective, and it should meet no more than six times per year, or the members almost become insiders and lose objectivity. Three to four times per year is effective unless numerous changes are going on within the organization or its environment. Five to seven members make up a common board size for companies with revenues of $10 million or more. Some prefer to start with fewer members until they see how it works. A board of three members can work as long as two of them are outside independent board members, meaning they are not employees or service providers to the firm. While an odd number of members is standard, there is more discussion than voting on boards of closely held companies, so an odd number is not imperative.

Closely held companies or family businesses have some unique problems that do not seem to affect businesses where ownership and top management are separate or where top management does not own a controlling interest in the company. In closely held companies and family businesses, where the roles of the owner and manager are usually performed by the same people, selecting an active, independent board of advisors or directors becomes a critically important decision. It brings

Figure 2.4 A board of advisors can be an important business tool for any company.

in objectivity so vital to the leadership of closely held companies. The board can positively impact a construction company's future prosperity by helping owners manage their business risks and gain control of their two-hat responsibilities. The owner and key managers then play critical roles in planning for the organization's business success in line with the board's guidance.

Review Questions. Check all that apply.

1   A 2012 contractors survey reported:
   a   As company grew bigger revenue per employee increased
   b   As company grew bigger revenue per employee stayed the same
   c   As company grew bigger revenue per employee decreased
   d   None of the above

2   Important functions of a business owner are?
   a   Establishing short-range and long-range goals
   b   Reviewing and recommending policies
   c   Measuring the performance of management
   d   All of the above

3   Boards of advisors may not be effective if it consists of:
   a   Only company managers
   b   Only people closely associated with the business
   c   Only family members
   d   All of the above

4   Which of these is a preferred board member?
   a   Someone very knowledgeable about construction
   b   A personal friend
   c   A person with "profit and loss" responsibility in a business
   d   An owner or senior manager of a successful business

5   For closely held companies or family businesses, an independent board of advisors is:
   a   Not necessary
   b   Critically important
   c   Optional
   d   None of the above

Critical Thinking and Discussion Questions

1   Describe the difference between a functional analysis and an organizational chart.
2   Explain why some organizations that outgrow their workload become top heavy.
3   Describe the information provided by a typical organizational chart.
4   Explain how responsibilities of the owners of a construction business differ from those of management.
5   Describe the function of a board of advisors.

3   Business Planning

Long-term business planning is essential to the continuing success of a construction enterprise. A business plan is typically presented in a document that discusses a set of management decisions that will result in the success of a firm. In its most basic form, business planning is the process of deciding what a firm will do to achieve success, and how it will do it. It helps provide a logical and rational sense of direction for a company. It also has internal and external uses. Internally, it can help improve performance by identifying both the strengths and weaknesses of the company's operation, and potential or emerging problem areas. It should communicate to management and staff clear expectations regarding the company's performance and priorities and provide a solid base for measuring overall performance as well as that of individual units and managers. As new developments and opportunities arise, a business plan provides a rational structure for evaluating the impact of the developments or opportunities on the operations and performance of the entire firm and on individual functions.

Perhaps most importantly, a business plan, and the process of developing it, can be used to educate and motivate the managers and key staff of a company. By analyzing past performance, evaluating the impact of trends and developments, and putting together action plans for the future, managers and other key staff tend to learn about the total operation of the company and the relationship of their particular area of responsibility to the achievement of the firm's goals or success. This level of involvement and understanding almost always leads to increased commitment to achieving the company's goals and often generates higher levels of motivation.

In fact, business plans can play a role in internally inspiring the organization. Therefore, the authors recommend drafting a business plan with a positive flavor and easy-to-read style, so that key people in an organization can easily relate to the plan. No matter what the size of the company or stage of development, success lies in getting the key people to pull together and get excited about what is right for the organization and their important role in it.

A business plan also has external uses with outside parties seeking financial and non-financial information. It can be used to educate partners like the bank or bonding company about objectives, structure, and performance. With this use in mind, the authors have included a sample business plan in this chapter,

DOI: 10.1201/9781003229599-4

containing information that outside parties are interested in. A good business plan provides clarity, both internally and externally, and should be easy to read and understand.

The authors want to emphasize the planning process itself is as important as the outcome. In the next section, the authors discuss practical approaches that contractors can consider for successful business planning within their organizations.

Practical Business Planning for Contractors

Throughout history, individuals engaged in business have always planned. From ancient hieroglyphics we know of plans for the great structures of our earlier counterparts. Even today, we certainly would not attempt the construction of a bridge, building ventilation system, or pipeline without preconceived drawings, a schedule for the timing, labor loading, and procurement of work. With the amount of time and effort that contractors put into planning work, and as familiar as we are with the necessity and value of good planning, you would expect we would be among the first to understand and utilize comprehensive business planning in managing our business enterprises, not just our projects. It is simply not yet the case in the construction industry. Although larger construction firms often have detailed business plans, smaller and medium-sized contractors often do not. Given that small and medium-sized firms make up the majority of construction companies, this lack of business planning significantly impacts our industry as a whole.

The process of planning may be defined as deciding in advance what is to be done, when it is to be done, how it is to be done, and who is to do it. In addition to identifying future opportunities and threats to be exploited or avoided, effective planning provides a framework for better decision-making throughout the company. Corporate planning is also an effective defensive tool against business threats, which alone is an important enough reason for smaller businesses to embrace it.

A business plan provides guidance to the managers of a company for making decisions in line with the goals and strategies of top management. It helps prevent piecemeal decisions and provides a forum to test the value judgments of decision makers within the organization. Perhaps the most significant value of a well-organized planning process is the improvement in communications among all levels of management about goals, objectives, strategies to achieve them, and detailed operational plans.

Planning is not making future decisions. Planning is concerned with making current decisions in light of the anticipated future (Figure 3.1). It is not what should be done in the future, but rather what should be done now to achieve outcomes in the uncertain future. Decisions can only be made *in* the present. Yet, decisions cannot be made only *for* the present. Once made, the strategic decisions may have long-term irrevocable consequences.

When properly performed, the planning process creates a communication network within even the smallest of companies that gets people excited about what's right for the company and how to achieve it. Planning also addresses an area sadly lacking in most small businesses today: the "measurement of success." Establishing a fundamental level of corporate planning in smaller contracting businesses has profound effects on the outlook, attitude, and performance of employees and business owners alike.

Many in the construction business are so involved in the planning of project work that they are distracted from, or lose sight of, the importance of planning for the business entity itself. In an informal study by the authors, looking at several thousand contractors of all sizes and types throughout the US over more than a decade, less than 40% claimed to do any type of formal corporate planning at all, and many who said they planned did not formalize their planning process in writing. This finding was confirmed in a formal study of specialty contractors, also conducted by the authors, showing only 37% of contractors have a written strategic plan (El Asmar et al. 2017).[1] That percentage demonstrates that a lot of companies still do not acknowledge that long-term business planning is essential to the continuing success of their construction enterprise. Quoting Yogi Berra, "*If you don't know where you are going, you'll end up someplace else.*" Planning is the ultimate risk control tool for the business.

Another informal survey of middle managers generated some disheartening information. One-third of the middle managers whose companies did plan indicated that while their company had a formal written business plan, the organization did not follow it closely or the contractor changed direction from the plan without notice. Not following a plan is the same as not having one, although many respondents said that the annual planning meeting had value even if the plan was not followed.

One of the most common reasons for not planning is company size. Believing that a smaller business does not benefit from a plan or that planning has little advantage for a midsize or a larger well-organized company is a serious business

Figure 3.1 Planning is concerned with making current decisions in light of the anticipated future.

error. Smaller companies often have limited resources and cannot afford a trial-and-error approach. There are over a million separate contracting companies in the United States, and the largest percentage of these comprises small businesses. In this competitive industry, many contractors succeed by personally and aggressively driving their business forward. However, the importance of effective and efficient comprehensive corporate planning and its impact on the success or failure rate of construction businesses have been understated within the industry. It has been underestimated by contractors for too long. And while more and more companies are embracing business planning, the construction industry cannot count itself among mature and sophisticated industries until that number approaches 100%.

As discussed briefly earlier in this book, the three primary functions of a construction business are getting the work, doing the work, and accounting for the work. Once a construction business is broken down into these functions, and time and energies are budgeted to treat them separately and properly, the company is fairly organized to properly manage the day-to-day activities of the business.

It is also important to address the responsibility of top management for the longer-range goals of the company, which dictate the well-being and success of the business. Properly managing the important day-to-day marketing, production, and administration areas of the business is not enough to ensure success; short-term success in no way implies long-term prosperity.

Without some forecasting and planning, businesses can be driven in the wrong direction. To carry the driving analogy further, contracting businesses have no reverse gear. If someone drives too far in the wrong direction, they cannot simply back-up and restart. Once the company and its resources are committed in a certain direction, it generates momentum. Changing that direction can be difficult, expensive, and maybe too late.

A great selling point for comprehensive corporate planning is the ability of the contractor to simulate the future, on paper. If the simulation doesn't work out, the exercise can be erased and started again (Figure 3.2). The mere ability to experiment with different courses of action without actually committing resources encourages the participants in the process to stretch their creative skills in a safe environment. Models of real-world situations provide an opportunity to test different scenarios and their possible consequences. In an exercise, decisions are reversible. Ideas may be tested without committing resources to them or betting the entire company on them. Simulating various business scenarios encourages and permits management to evaluate many alternate courses of action. This could not happen in the real market. Additionally, the simulation process supports the contractor in selecting the "right" course of action as it becomes more apparent. There is also the possibility that the larger number of alternatives may produce ideas that would otherwise have been missed. Corporate planning allows top management to better anticipate new opportunities with greater lead time. With more notice and a predetermined course of action, exploiting new opportunities prior to the competition is much more likely. Another side to this coin, and equally important, is being able to look ahead to reveal threats to the business before they unexpectedly materialize.

Figure 3.2 If the simulation doesn't work out, the exercise can be erased and started again.

While no one can predict the future 100%, the probability of certain events having a predictable cause and effect relationship is pretty good. The more we know about our business, our market, and our competition, the more likely we are to simulate the outcome of our moves with increased accuracy. In addition, the more we plan, the better planners we become. In preparation for planning, readers may find it helpful to use the *Construction Company Self Analysis Program* developed by the authors. The program will provide insight into the strengths and weaknesses of the firm. It is available at no cost on this book's companion website: www.routledge.com/9781032134734

The program is located under *Support Material* toward the bottom of the page. Note that this tool works best on MS Windows machines. In the next section of this chapter, the authors will provide a sample business plan to help the reader get started developing, or revising, their construction company's business plan.

A Sample Business Plan

There is no hard and fast rule defining how to present a planning document. The plan and the plan document should reflect the size and stage of development of the construction company preparing it. A small organization might spend only five or ten hours planning, which may be presented in a few pages or bullet points about what is expected of key individuals and what is intended to be accomplished. A large organization may have a much larger document broken down into divisions and departments. Midsize companies should have as much detail as the planning group feels necessary to capture what they intend to do and who is to accomplish what and by when. The important thing is the plan should reflect the planning group's decisions and be understood by everyone in the organization. The plan is what counts; the document simply reflects that.

A planning group is essential to outline and drive an organization's business plan forward. In most organizations the planning group will be the key personnel who run the organization every day. The size of the planning group generally varies by the size of the organization, but the ideal group size is between 3 to 12 key people. Groups including more than 12 members can be difficult to manage. In a smaller company, three or four key people may be enough for forming the planning group. In a larger company, all department and/or division heads may be reasonable candidates. The ideal planning group has representation from each functional area of the company (i.e., one person from accounting, one from estimating, sales, etc.).

The qualifications for group members include being strategic and forward-thinking individuals. Some people are great in considering the here and now and even the near-term future but have limited ability to conceptualize the long-term future. The selection can be complicated in that planning requires strategic thinking and you may sense that some long-term or senior employees believe they are "entitled" to participate or have earned the "right" to serve but are not as qualified as others. The quality of the plan is directly proportional to the qualifications of the planning group members, and so hard choices may need to be made. Compromising by adding "deserving" less-qualified people does not work and will diminish the quality of discussions, deliberations, and decisions. Facing the hard decisions upfront is far superior than allowing the quality of the plan to suffer. The planning process should be transparent, so it is appropriate to keep all permanent employees advised of the pending process. A memo to employees, introducing the planning process, should help explain to those not selected how the process will work (Figure 3.3).

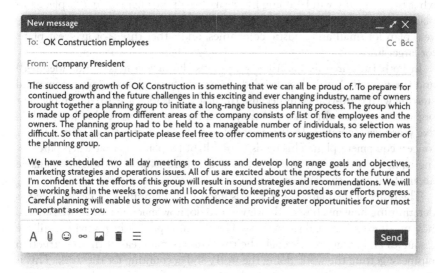

Figure 3.3 Sample announcement memo to employees.

Company owners and non-owners are treated as equal participants during the planning sessions. However, in some cases, personality conflicts inhibit smooth planning sessions, and some organizations do better with a facilitator to enable the planning process and write up the results. It must also be noted that facilitators should carefully limit their own input in the planning process to avoid influencing outcomes. A minimum of one key field personnel is imperative in the planning group; however, it may be best to include additional field representatives to ensure that the planning group is not too office-centered. In the authors' experience, contractors that remain successful through good times and bad times are field-centered construction companies, not office-centered.

It is imperative that planning sessions be held away from the office for several reasons. It is very difficult to get employees to act as equals when several levels of management are present. It can be uncomfortable for employees at the bottom of the reporting chain to speak their mind to those of a senior rank, but their input usually has huge value as they often know more about what is going on in the rank and file than top management. Being away from the office demonstrates that this activity is different and helps validate that in this setting everyone's opinion is valued and accepted as equal to anyone else's. There are often "strong willed" in top management to the extent that in some cases a facilitator will be of great assistance in getting the best out of everyone.

You should plan for your planning horizon which is as far into the future as you can project your market. For a small or midsize construction enterprise, a three-year plan is appropriate because it is difficult to see forecasts much farther out than that. Planning year one can be reasonably accurate because backlog and available market are fairly clear. The second year is harder, and by three years out market projections are primarily guesswork. Some recommend a five-year plan, and we are not strongly opposed to it. It is just that putting effort into what the company will be doing five years from now might be better placed in planning on how to get there in one-year increments. We recommend a three-year plan for small and midsize companies, and either three- or five-year plans for larger firms.

In both the three-year and five-year plans the first year is "hard" (detailed) and the next are "soft" (less detailed). However, for goal-setting we recommend starting with third-year goals because the first year is too close to "dream big." For organizations that own a lot of heavy equipment with a useful life longer than five years, we recommend a three-year plan with an attachment of a five-year (or longer) equipment plan. This tends to highlight for planners that when purchasing equipment that lasts five or seven years, there is little knowledge available about the market the equipment is to be working in after three years.

The order of the planning process is important. Planning can be described as determining how much work do we want to do, how much work can we get, and how much work can we do for a profit. The planning process begins with available resources and progresses through the three functional areas of the construction business: getting the work, doing the work, and accounting for the work. This is a logical sequence that requires the least amount of rework which is an issue in any planning effort. The segments of the plan are interactive, and the completed segments must be subject to change as subsequent portions are developed and available resources come into question. The owners' and planners' goals and objectives should be clarified first (more about this later). Then a realistic projection of the company's market over the years, the business plan will cover and the development of a marketing plan on how the work will be captured.

An organization or operation plan is next, which simply describes how the company will be (or already is) organized and structured to carry on the work of the enterprise and how that will be accomplished. The planners should clearly depict the reporting relationships of the key people in the organization and chain of command. Many small and midsize construction companies lack this fundamental management tool that leads to misunderstandings and confusion in the ranks. The longest portion of a first-time planning effort is often devoted to the activities the organization needs to accomplish to produce work at a profit. The last plan to be developed is the financial plan. The financial plan translates all the other segments of the plan into dollars. It shows what the company's income statement will look like in each of the plan years if the organization accomplishes the plan goals, objectives, and activities. A rough pro forma income statement can be used as a measuring stick.

The financial plan is actually one of the softer (less detailed) portions of the overall plan as it translates into numbers that the planners hope to accomplish, which often consist of ranges, and as such numerical results cannot be predicted with great accuracy. In some cases, the financial plan may translate the rest of the plan in a negative light, meaning that the results of the other plans will not produce a satisfactory picture. New equipment may have been scheduled to be purchased in the operations plan that creates an unsatisfactory debt ratio in the financial plan, or sales may be planned for in subsequent years that require bonding capacity that the pro forma statement will not support. As with the other stages in the plan, if the current plan is not compatible with the previous segments of the plan, then the previous plans needs to be adjusted. By following this sequence, planners will be able to develop a practical, resource-based business plan that will work for the company. The plan is a tool to manage with and generate excitement throughout the organization. If first-year goals are achieved, great! If not, the planners will be able to dissect, discuss, and discover the reasons for that, and will be better prepared to set realistic goals and plans for the following year. The planning process is not only a great tool but also a great teacher.

The following is a sample of what might be included in a written business plan. It is fairly inclusive for teaching purposes, so the reader may use any parts that apply to their circumstance. The style of the writing is not important, and the sample is simply one way of doing it. Just using bullet points that the planners understand is also a reasonable approach, but may not be as easily understood by outsiders, such as banks or bonding companies.

A business plan for a hypothetical construction company called OK Construction is discussed. In this example scenario, the owners of OK Construction decided to initiate formal long-range planning as a method of managing their business. *Owner 1* is president of the company and its chief executive officer. *Owner 2* is responsible for marketing and equipment management. *Owner 3* is in the process of assuming greater responsibilities for construction operations and will be responsible for field operations in three years.

First, a selection of key individuals representing all functional areas of the company established the planning group and began the planning process. A three-year plan was developed, including a first-year plan in greater detail. The plan covered marketing, organization, production, and finance, with short-term and long-term goals and objectives established in each area. To accomplish the goals and objectives, the group formulated strategies and scheduled specific actions. The planning group met quarterly to monitor progress and adjust the strategies to current situations.

The owners and the planning group are extremely pleased with the results of the planning process after one year. While volume goals were not met, the percentage of gross profit exceeded the planned amount. The planning process helped focus the attention of the entire company on the bottom line. Overall corporate performance was excellent, and the current market looks favorable for continued growth and success.

After one year, a new three-year plan is developed by the same planning group; again one year in detail and two years with less detail. The exercise is repeated every year, looking three years in the future. Several activities or objectives planned in the original schedule, but not accomplished from the previous year, were restated for action the coming year and new objectives and strategies were added.

The new one-year plan, while like the previous one, is unique to this plan year and based on a careful evaluation of previous year's performance and results. The plan development was completed in a shorter time span than the previous one because the group had planning experience and an existing plan to modify.

In this scenario, significant elements of the first-year plan included:

- The planning group elected a new member, expanding the field representation to three, with the addition of *Employee a.*
- *Owner 3* will work with *Employee b,* and by the end of the third planned year, *Employee b* will be responsible for field operations.
- The accounting systems will be integrated by the beginning of the next fiscal year, so monthly financial statements can be computer-generated.
- *Owner 2* will establish a better system of dealing with the billings for co-owned equipment from a subsidiary company, and evaluate both companies' performance.
- *Owner 2* will establish an equipment management system to schedule replacements as needed for growth or obsolescence and to assist with the scheduling of future financial needs.

A table of contents for the sample business plan of OK Construction is shown in Figure 3.4. This table of contents covers a broad spectrum of subjects as an illustration; not all of these topics may be necessary for every company. The typical planning group would normally have fewer subjects covering topics the company

Figure 3.4 Table of content of sample "catch-all" business plan.

is having issues with or has an interest in for any number of reasons. Therefore, the following samples of each subject depict a longer plan than a small or midsize company would likely have. The following sections will delve into each of the headings identified in the table of contents.

I Executive Summary

1   The Company and Its Environment

OK Construction is a building construction company working in three states with home office and shop facilities centrally located in-state. The company owns and operates its heavy equipment, including cranes and excavating equipment. Health care facilities, industrial projects, commercial buildings, and school construction are the primary markets. Jobs range in size from $1 million to $10 million, with smaller work taken closer to the company location.

The company is owned and managed by *Owner 1, Owner 2,* and *Owner 3.* Annual volume has been steady for two years, in excess of $30 million, with profits at industry standards. With the current organization and equipment, the company believes it could eventually produce $40 million annual volume with little or no increase in overhead. (Author's note: This would not be easy to accomplish without an increase in overhead resources which will add costs.)

The client base is made up of large industrial clients; private, commercial, and institutional owners; and public bodies such as school boards. Heavy industrial and complicated institutional and commercial projects are the primary target utilizing the company-owned cranes and equipment. That type of work, while competitive, appears to be available for the plan period.

2   **Goals: Non-financial and Financial**

The owners intend to operate aggressively by selecting, purchasing, and producing to maximize profits and minimize exposure in the challenging years to come. They control their business future by structuring and managing a highly efficient organization that fully employs the existing resources and is flexible enough to respond quickly to changes in the market. Structured and controlled overhead is employed to minimize the effect of unforeseen downturns in the economy.

A growth to $39 million in annual volume in three years is intended with profitability over 3.5% after taxes, and a 10% annual increase in corporate net worth. Replacement of equipment to remain competitive is planned, as needed, with little increase in the number of machines during the plan period.

3   **Strategies**

By the third year, a service mix of 40% health care facilities, 40% industrial work, 10% commercial, and 10% institutional projects is planned.

Work will continue to be obtained through negotiations as well as the competitive bidding process on select jobs, with approximately 50% of each. *Owner 1* will put greater emphasis on soliciting new industrial clients. *Owner 1*'s targeted selection efforts (for both projects and clients) are intended to strengthen competitive advantage.

Favorable banking and bonding relationships exist and should continue through the period particularly with the planned growth in net worth. Maintenance of existing clients will continue, with *Owner 2* co-ordinating the efforts.

The owners have solidified their organization and increased profits by carefully managing their work. Continued improvements in marketing and bidding strategies, equipment and purchasing management, and production controls will allow them to continue to operate profitably with less risk. Careful planning has positioned the company for future growth if the market expands, while allowing for safe retraction if the market declines. The owners believe that careful planning prepares them to operate profitably if the market increases or decreases.

II Marketing Plan

1 Marketing Philosophy

OK Construction has been serving their market for two generations and enjoys an excellent reputation for quality construction on schedule. The company has earned the respect of architects, engineers, and owners by providing their services in a highly professional manner. It is client-centered.

2 Sales and Revenue Objectives

The owners believe the market will develop favorably over the next three years. Sale goals are $34 million for the first plan year, $37 million in the second year, and $39 million in the third year. Profit objectives are 7%, before general and administrative expenses.

3 Product Line Strategies

Anticipating an increase in competition in the commercial and institutional markets, a change in service mix is warranted; 40% industrial, 30% health care projects, and 30% commercial and institutional buildings will be sought. The company's primary geographic area includes the surrounding states which will not change during the planned period.

3a Target Projects: Projects to pursue will be categorized by type of work rather than by client. Continued emphasis will be placed on selecting projects to fully utilize company-owned equipment. Additional local industrial projects will be sought within a 40-mile radius of company location because of the profit potential from these jobs and the fact that they can be easily managed from the home office. Projects over $5 million will be targeted, with consideration given to the availability of superintendents qualified for that particular work.

3b Marketing Analysis: Study of available work will continue so that projects which are of particular interest to the company can be targeted and a focused marketing approach can be developed to capture them. *VP Marketing* will continue to prospect for projects in the planning stage and pursue targeted industrial clients within a 40-mile radius. This will enable management to know, with greater lead time, what projects will be available to bid on and to make selections in accordance with desired service mix and available superintendents. It will also be used to identify project opportunities for negotiations which may otherwise have been put out for bid.

3c Bid Strategy: A more formal method of identifying primary and secondary competitors and tracing their work loads and need for work will be developed in the first year and refined in the second and third years. Evaluating the competition's need for work at a given point in time helps measure competitive pressure on projects being bid and can improve success ratio and narrow bid spreads. Data collection will begin immediately for use in an organized manner next winter. The *President*, *VP Marketing*, and *Employee b* will be involved in the selection of projects to be bid and will work with the estimators on pricing. Final pricing will be based on current need for particular work type, the number of similar projects available, and an appraisal of the other bidders' perceived need for the job. Project selection and pricing will be driven by company needs and requirements to the extent that choices are available.

4   Marketing Activities

The marketing department conducts a number of activities to keep the company in the minds of prospective clients in a positive way.

4a   **Newsletter**: The company newsletter is sent to clients and prospects three times a year. The mailing list is managed by the Marketing Department, and all key employees are encouraged to submit names they consider appropriate.

4b   **Equipment and Job Signage**: Equipment and job site signs are considered an integral part of the company's image. All trucks, office trailers, and larger company-owned equipment exposed to the public view will be similarly lettered with "OK Construction" and kept reasonably clean and neat. Signs on all projects where permitted and appropriate will be displayed with "OK Construction."

4c   **Press Releases**: Press releases are prepared and distributed to appropriate publications on all items considered newsworthy that place the company or its key personnel in a positive light. To maximize these efforts, suggestions from key employees are sought.

5   Current Year Marketing Strategies

To capture the number and size of projects within the predetermined service mix, management targets specific projects and the broader categories of work and potential customers. *VP Marketing* is responsible for getting the work. The *President* will spend a minimum of 20% of his time and *VP Marketing* will spend 100% of her time marketing.

5a   **Recording Contacts**: A system of recording all marketing contacts by phone, mail, or in-person has been developed. Information about the potential client's needs, his or her likes and dislikes, and significant comments that the prospect put in writing will be captured. This will facilitate the coordination of repeat calls with a client, which in some cases will be made by different people and even at different levels within the organization.

5b   **Repeat Business**: Repeat business is considered a primary source of work. A systematic method to contact clients, which the company has worked for in the last ten years, has been undertaken and will continue.

5c   **New Industrial Clients**: To increase the company's market share of industrial work in the area, a list of 40 potential clients within a

40-mile radius has been developed and a number of contacts made. This list will be refined to 10 or 15 good prospects that will be intensely pursued by *VP Marketing*. The initial jobs may be smaller, and success may require repeated efforts for a year or more; however, two to five new industrial clients are anticipated within 18 months.

5d Client Maintenance: Specific client maintenance strategies have been developed to maximize repeat business from existing clients. Marketing at more than one level within the client's organization is performed where appropriate. Specific activities are agreed upon including who will carry them out, how often, and who will coordinate the effort. Client maintenance programs have been developed for large *Client a*, *Client b*, and *Client c*, and others will be added as appropriate.

5e Target Projects: Target projects of particular interest to the company are regularly isolated. A detailed strategy is developed for each, outlining exactly what activities will be undertaken by whom, and when to maximize the probability that the company will get the job. Target projects will be added to the list as they develop and all key people within the company should know about them. Anyone hearing anything about a target project should inform the *VP Marketing* about it.

III Organization and Management Plan

1 General Philosophy

The owners of this closely held company have a high-level of concern and respect for their key employees and feel strongly that this group of people and the company will grow together. This nucleus has the potential to manage the $34-million target volume and more. The group is managed in an atmosphere of mutual respect, and informal lines of communication are encouraged. The current comfortable working environment will be maintained, but growth and efficiency will require a more formalized reporting chain to be developed. There is accountability at all levels of the organization.

2 Organization Structure

The organizational structure of the company is shown visually, with reporting chain and lines of authority (*refer to Chapter 2 for sample organizational structures*). The Marketing and Estimating Departments report to *VP Marketing*. Field Operations and Equipment report to *Employee b*. Field superintendents are responsible for labor costs, safety, and training and report to the project managers. The project manager is responsible for the profitability and success of the job. On technical issues, the superintendents and project managers may receive guidance directly from *Employee b* who will be spending 20% of his or her time on operation and production. *Owner 3* will assist *Employee b* full-time in production and field operations. He or she will assume more responsibility during the three-year plan period, with a target of full field responsibility by the end of the plan period. A specific course of action will be developed by the *President* and *Employee b*, including seminars and equipment demonstrations to be attended and learning activities to accelerate the training process.

3   **Politics and Objectives**

The owners will continue to encourage the individual initiative of key personnel within their areas of authority and remain open to suggestions and recommendations on all subjects. It is the collective knowledge of our key people in construction, bidding, and administration that makes OK Construction the dynamically successful organization that it is. Providing for the continued growth of its personnel in a safe, healthy, and stimulating working environment is an important ingredient to the future of the company.

4   **Training and Development**

All employees are encouraged to improve their skills with on-the-job training and working with others who may know more about particular aspects of the business or possess skill levels greater than their own. Individuals who feel they can do more or wish to advance within the company should make it known to the owners. Independent training or education in skills and knowledge directly related to the employee's job or advancement is encouraged and on a case-by-case basis may be paid for by the company. Periodic related work and personal development opportunities will be available. Group learning and social events throughout the year will provide opportunities for employees to exchange ideas and interact with middle and top managers on an informal basis.

The identification and training of potential foremen, field superintendents, and project managers is an important aspect of the company's future success. Key employees are encouraged to point out capable employees to be considered for advancement and try to determine their level of interest in moving up within the company. After discussions with the appropriate managers, the selected candidates will understand that they are being considered and will be assigned to work under key company people who have agreed to teach and train them. Scheduled reviews keep the process on track. At those review sessions, outside reading or specific instruction or training may be recommended. The candidates will be encouraged to understand that the company is willing to invest in them if they are willing to invest themselves in the company. One new project manager will be hired during the first year of the plan.

5   **Compensation**

The company's union workforce is compensated in accordance with local agreements. Salaried personnel are paid locally competitive wages to maintain the best available people. Management staff are compensated for their knowledge, efforts, and loyalty on an individual basis, and the owners concern themselves directly with this issue. The *President* will review each key person annually and discuss the subject personally and frankly with them. Increases are based on merit and subject of salaries should be treated confidentially.

IV Operation/Production Plan

1 Production Scheduling

Employee b, in consultation with others, will schedule all construction operations and assign owned and rental equipment to projects. Planned production rates and methods will be discussed and agreed upon with the on-site project superintendent and the project manager, and then measured. Meeting or exceeding planned production rates defines the entire field organization's focus. Every employee, whether moving a piece of equipment or managing the largest project, needs to remember that his or her actions, as part of the team effort, can make the difference between profit and loss of a work phase or an entire project. The company relies, exclusively, on field operations for its cash flow and profits and on the efforts of each employee. This message will be highlighted at all employee meetings and promoted on a continuous basis by all management personnel.

2 Production Standards and Costs

OK Construction will maintain its high standards that foster its reputation for high-quality work on schedule. Further, the cost of rework or removal of substandard work makes high-quality standards the most efficient and economic approach to produce construction projects at a profit. The current methods of recording and measuring production costs will continue for the first year of the plan but may change in year two because of changes being made in the accounting department. All employees should concern themselves with the accurate reporting of requested information on time, and field supervisors are relied upon for the accuracy of the data reported. Management must receive accurate project information continuously to effectively control the work and provide the resources necessary to optimize field production.

3 Operating Policies

In this section, plans relating to the operations-side of the business will be detailed. This includes planning for purchasing, subcontracting, equipment maintenance, equipment management, and facilities.

3a Purchasing and Subcontracting: *Employee b* has overall responsibility for purchasing and subcontracting, while project managers will handle all purchasing on their jobs in consultation with him. Field supervisors may purchase minor miscellaneous supplies but should clear major purchases with the office. Common sense is to be used when any purchase is necessary to keep a job moving.

3b Equipment Maintenance: *Employee b* is responsible for equipment maintenance throughout the company. Field maintenance is a project responsibility but will be carried out in accordance with schedules and criteria established by shop management. Schedules for the renewal, overhaul, or upgrading of equipment will be prepared for a three-year period, with a majority of critical work done in the winter. Supplies and spare part inventories will be planned and maintained by shop management to minimize down time exposure, particularly at peak periods. The proper operation and maintenance of equipment is an important cost factor, and every employee is expected to understand their part. This will be stressed at job sites and safety meetings.

3c Equipment Management: During the first year an equipment management plan will be developed for integration with the maintenance procedures in which the useful life of each piece of equipment will be considered. The projection and control of lifetime equipment costs provides an additional tool in selecting projects most compatible with company equipment, and in the selection of equipment suitable to changes in future markets. An overall equipment management plan is consistent with the owners' desire to anticipate future financial needs.

3d Facilities: The company's home office and shop facilities are capable of accommodating growth in excess of the plan period. No changes are anticipated.

4 Estimating

Marketing is responsible for the estimating department. To enhance efficiency, certain activities are identified and targeted during the plan period.

4a Price Catalog: The unit price catalog will be updated at least annually.

4b Estimating Department: The possibility of establishing a separate estimating department will be explored during the first plan year.

4c Quantity Takeoffs: A target of zero errors in quantity takeoffs is strived for, and estimators must be alerted to special situations, such as areas where a machine or crane cannot reach or loads too heavy or too far. The constructability of a project should be considered by everyone working on an estimate.

4d Pricing Reviews: The estimators will attend all pricing review meetings for training and input.

4e Backup for Employee b: During the plan period, the company owners will search for a backup person for *Employee b's* position. By the end of the plan period, an individual will be selected and begin training to eventually lead the department.

5 **Productivity Strategies**

To achieve the company's overall goals, the gross margin or job profit, before general and administrative expenses, is paramount. A target of 7% has been set by the planning group and a number of practical activities have been planned.

5a **Superintendent's Field Time**: Field superintendents will spend a minimum of three hours per day outside of the field office directly supervising the work. To the extent paperwork, subcontractor coordination, or other activities interfere with this policy, they are to notify management for immediate resolution of the conflict because the planning group is convinced that the profit is made or lost in the field.

5b **Safety**: The superintendent as the senior full-time representative of the company in the field is responsible for safety. The company is dedicated to providing a safe working environment and targeting zero accidents on all projects. The superintendent must be aware of all aspects of the company's safety policies and procedures. Regular safety inspections will be conducted by the firm's insurance company and by the safety director of the firm's trade association.

5c **Crew Size Analysis**: The superintendents will analyze crew sizes, particularly on major activities, before starting the work and periodically, during each activity to maximize output. If the crew size or makeup is changed from the estimated makeup, it should be to improve productivity and the results should be reported to *Employee b* for possible future use. The estimate is based on crew sizes historically proven to produce a known quantity of work at a certain price. To ensure a successful project within estimated cost, the superintendents must meet or exceed our estimated labor costs. Management should be advised immediately if cost estimates are not met.

5d **Crew Communication and Goals**: To manage labor and maximize productivity, the superintendent should communicate to the field force what the short- and long-term schedule is and what goals and objectives he is trying to meet. The crews should know what they are trying to accomplish, by when, and how that fits into the overall project. Most people prefer to be involved in a successful project, and successful projects make money. Making a fair profit on the work should be discussed with field labor as a measure of success. "Profit's not a dirty word"; it is what keeps their industry going.

5e **Record-Keeping**: The superintendent should understand the need for appropriate record in the field; therefore, he or she should provide accurate and timely information and reports as requested. The non-performance of subcontractors or suppliers should be noted on daily reports and the project manager advised so prompt action can be taken.

5f   Subcontractor Management: Subcontractor performance can critically impact a project, and the management of subcontractors is primarily the project manager's responsibility. It begins with the selection of the subcontractors and the care taken to get the best available contractors for the work. This isn't always possible in a competitive world, and to remain competitive, a subcontractor's price is of paramount importance. The project manager should find out as much as he can about a new subcontractor by asking for a list of previous contractors he has worked for and making direct inquiries. Subcontractors with successful relations with the company on prior projects will require less attention, compared to subcontractors with questionable reviews. Considering project time, the company must carefully monitor the performance of questionable subcontractors until proven as a proficient team member. Financial information should be gathered on new subcontractors and any that are weak financially should be screened out. If for any reason one is engaged, they need to be carefully managed and early payments avoided. It is appropriate to get a breakdown of second- and third-tier subcontractors and suppliers from a financially weak subcontractor in advance of the work. Advising the subcontractor on proof of payment to each of these for the prior month will be required before payment is made, and, when questions arise, the project manager makes direct inquiries regarding payment or issues joint checks.

Performance of all subcontractors and suppliers is monitored closely, and written notice given to them for any area of nonperformance. As soon as it is believed that a subcontractor or supplier will delay a job or cost the company money, the project manager will notify senior management. Continued nonperformance will result in a meeting of senior people involved in the project to determine the appropriate action to take.

The project manager is responsible for getting the subcontractor to the project and ultimately for the performance of the subcontractor. The project manager should monitor performance through the superintendent and/or directly. Failure to report consistent nonperformance to management will not be accepted. The first issue in subcontractor management is getting the subcontractors to perform and keep them on notice of the project's needs and their response. The second issue is to get help when needed. Do not wait until a subcontractor has harmed the project before advising upper management. At a more senior level in the company, different actions can be taken. OK Construction leaderships must cultivate an attitude of success with or without buy-in from the subcontractors.

5g   Project Manager Time: Project managers will visit the projects they are responsible for at a minimum of once a week. The time spent in the field may vary with the stage of the project and coordination. Paperwork is a critical and important part of managing a project, but the work is built in the field, and the project manager is responsible for the construction of the work at a profit. As soon as it appears that isn't going to happen on any project, senior management must be notified.

Working as a team can bring the right talent to bear on a problem and turn it around before it gets serious, but only if we are skilled at spotting little problems early on and reacting as if they were already serious ones. In this business, there is no such thing as overreacting. It just seems like it sometimes when the serious problem is prevented from developing.

5h   Preplanning: Preplanning of all projects will take place prior to the start of the work. The project manager and superintendent with the help of *Employee b* will work out the details and schedule of each project. The project manager and superintendent will present them to management for discussion. The plans need to have some flexibility, but a greater level of commitment to the original plan is desired. Too often changes made in haste for short range gains are not advantageous in the long run. The project manager should analyze, carefully, any departure from the original plan and discuss any major changes before enacting them.

5i   Focus on Field: Descriptions of the reporting chain and levels of responsibility are intended to clarify who is responsible and accountable for building the work at a profit. Management focuses the company's energy directly on field operations and the production of successful and profitable projects. Everyone should recognize that accounting, marketing, estimating, and administrating are functions that revolve around and support the primary objective of the enterprise – building for a profit. This concentration on construction operations by everyone in the company can put the excitement back into our work.

6   Supervisory Personnel

Superintendents and project managers should recognize that identifying and training good foremen is critical to the long-range growth and success of the company.

6a   Advancement of Personnel: A list of staff considered for advancement has been developed, and a systematic review of their progress will be undertaken at least once a year. Additionally, project managers and superintendents should bring to the *President's* attention anyone that they consider highly valuable for advancement.

6b   Planning Personal Needs for the Future: By the end of the plan year one, the number of new foremen, superintendents, and project managers needed in the next several years will be established and compared to the number in training. If necessary, recruitment for new hires will be planned for to keep up with the demand for qualified construction people to run the work. In the second year of the plan, a more formal method of indoctrinating and training foreman, superintendents, and project managers will be explored.

V Financial Plan

1 Accounting Policies and Controls

The CFO is responsible for accounting. All record-keeping tying into the general ledger will be completed in accordance with generally accepted accounting principles, in cooperation with the firm's outside independent auditors who will report directly to the owners. New or expanded accounting controls are in place as a result of last year's plan. No additional accounting personnel are anticipated, and planned growth should be absorbed by automation.

Existing monthly management reports may be altered next year to be automatically adjusted for over and under billings. Alternatively, a monthly cost control system may be considered that would be independent of the general ledger and tied to the original bid quantities and prices. Selected work items are to be tracked and reported by field management, processed by the accounting department, and evaluated by management. This will allow management to get quick and accurate measurements of project performances, which, in turn, are earlier than the complete statement information, generated monthly.

2 Yearly Performance Statement Projections

This section details financial performance statements of the current year and estimates for the second and third years. A sample is shown in Table 3.1.

3 Current Year Activities

The short-term activities for improvements in the company's accounting functions are discussed in this section.

Table 3.1 Yearly Performance Statement Projections

	Current Year Performance	Estimated Second-Year Performance	Estimated Third-Year Performance
Construction revenue			
Construction cost			
Gross profit			
General and administration expenses			

3a **Quarterly Reports**: The CFO will produce quarterly financial statements for management until systems integration is accomplished.

3b **Systems Integration**: The accounting systems will be integrated by the beginning of the next fiscal year. A commercial off-the-shelf electronic billing software will be purchased and put into use immediately. Similarly, an electronic payroll system will be evaluated starting in September. If it is not selected, the payroll figures will be put into the general ledger by category on a weekly basis. Equipment costs will be applied to each job as they are incurred monthly. Over and under billings will be applied on a monthly basis.

3c **Outside Accounting Costs**: The target for outside accounting services is a cost reduction of 10% made possible by the new systems that have been or will shortly be put in place.

3d **Subsidiary Company Charges**: The handling of material, truck, auto, and equipment charges between the company and the subsidiary equipment company creates a certain amount of confusion for those not closely involved with the transactions. Clarification of the process or a new procedure will be undertaken by the *President* and CFO; with a target for resolution of the situation by the close of the fiscal year. Changing the fiscal year for the subsidiary equipment company to match OK Construction will be considered, so combined statements can be produced. This will enable management to better evaluate the cost of equipment ownership and its impact on profitability.

Communication: Announcing the Planning Process and Distributing the Business Plan

Historically, secrecy has been popular in the construction industry; however, transparency is a valuable asset in today's business environment. While it may not be preferable to share the business plan with your competition, due to confidentiality concerns, it must be asserted that there is very little a competitor could with your plan, short of seeing how well managed your company is. To not share the plan with employees is to forgo a huge motivational opportunity. Employees like to know the company they are with will be successful into the future. A multiyear business plan suggests the firm is well managed and knows where it is going and how it is going to get there.

Initially, the authors recommend that the planning process be announced internally, as in Figure 3.3, to generate excitement and set up the anticipation for the plan itself. The sample memo shown in Figure 3.5 is an example of what might be sent delivering the business plan document.

Once the plan is developed, it is strongly recommended that the plan be shared as broadly as management is comfortable with. Many exclude the financial section of the plan other than to their bank and bonding company. If you do not intend to share the plan, it is highly recommended that you draft a summary of

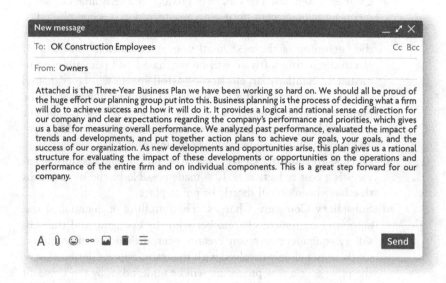

Figure 3.5 Sample memo to employees delivering business plan.

the plan, excluding any areas of concern, and circulate it to key employees or all employees. A memo summarizing the plan can be distributed to all permanent employees, perhaps included with payroll checks.

Summary on Business Planning

The days of assuming your business is helpless in the face of market forces are long gone. A substantial number of contractors today are realizing their businesses do not simply need to react to market-created booms and busts that have plagued the industry for so long but must plan for them. They are embracing an approach that suggests they can determine their future direction with proper planning. Through effective business planning, contractors can be assured that their established objectives are met, or they will at least know the exact reasons why not.

The construction industry is undergoing some dramatic changes, not the least of which is a realization among constructors that the "tried and true" old ways of running their contracting business will not suffice in this highly competitive and high-risk industry. There has always been a painful weeding-out process in the construction industry by which companies that don't keep up, fail. Others take their place, and some of them fail as well. Many assume this is simply a fact of life in a high-risk and high-stakes business. After many years of working with distressed and failing contracting businesses, the authors have determined the causes in a majority of the cases were management decisions. In the following chapter, the authors will detail the concept of flexible overheard, a valuable management practice for contractors in the business of construction.

Review Questions. Check all that apply.

1 Which best describes the size of the construction company that will benefit from a formal business plan?
   a   A large organization
   b   A midsize organization
   c   A small company
   d   Every size construction organization

2 A business plan:
   a   Describes what a firm will do to achieve success
   b   Describes how a firm will achieve success
   c   Provides a logical direction for the company
   d   All of the above

3 Which of these does not describe a typical business plan?
   a   A business plan helps identify future opportunities
   b   A business plan helps define specific construction processes
   c   A business plan is a defensive tool against business threats
   d   A business plan is a framework for decisions-making

4 Planning is concerned with making current decisions:
   a   Faster
   b   Just in time
   c   In light of anticipated future
   d   When required

5 Qualifications for planning group membership include:
   a   Strategic and forward thinking
   b   Having an ownership interest in the company
   c   Having worked at least five years with the company
   d   Having worked in more than one position in the organization

Critical Thinking and Discussion Questions

1 Describe the advantages of long-term business planning.
2 Discuss the common reason for not planning.
3 Discuss the qualification and process of selecting planning group members.
4 Explain how you would go about determining the appropriate length of a business plan.
5 How many members would you prefer on a business planning group and why.

Note

1 El Asmar, M., Sullivan, J., and Sullivan, K. (2017). Developing the New Horizons Foundation Decision-Making Framework for New Market Entry in the Sheet Metal Construction Industry: Research Report. New Horizons Foundation. Chantilly, VA.

4 Flexible Overhead

Introduction

The US economy has always been and continues to be cyclical. The construction market follows and lags the US economy. Therefore, the construction market as an integral part of the economy will continue to be cyclical. When the construction market is growing the industry prospers; however, when the market declines, the industry suffers reductions in profits and exposure to loss. The reason is that growth requires additional overhead expenses in the form of personnel and resources which are brought on as permanent fixed costs. When the market declines, it is extremely difficult to reduce fixed overhead because you cannot eliminate half a person or half a truck. The current construction business model of permanent fixed overhead is the cause of the industry's boom and bust cycle. This raises the following question: If a firm earns a certain percentage of profit putting X amount of work in place in a year, shouldn't it earn the same percentage of profit the following year when it puts in place, perhaps 15% less work? Most don't.

Construction enterprises make more money in good markets than they make or lose in downturns; however, because the length of growth periods has been longer than the downturns, most come out ahead. This has been going on for so long now that it is generally accepted as a cost of doing business. However, the cyclical nature of the business is one of the reasons the construction industry has the second highest failure rate in the nation. The business model must be modified to a process that allows construction organizations to prosper in both good markets and bad. It makes no sense that a growing company that has been profitable at $8, then $9, then $10 million in sales would automatically lose money if it is forced back to $9 or $8 million by a decline in the market.

If a construction company makes money growing to $10 million profit, it usually suffers if it is forced to downsize. To be profitable after reduced sales of 10% or 20%, they would have to immediately reduce overhead costs by a similar amount. Most say: "Impossible! It can't be done," and they are correct if overhead remains the same. Whether they recognize it or not, if a construction enterprise is forced to reduced sales they are also "forced" (if they choose to make a profit) to reduce overhead expenses accordingly. I say "choose" because this is a management decision and they have a choice, but reducing overhead is painstaking. *Flexible*

DOI: 10.1201/9781003229599-5

overhead is a process that allows the reduction of overhead costs without the stress of laying off core key people and incurring the cost of disposing of idled resources, both of which will be needed in the future when the market recovers.

Planning on continuous growth cannot work in a cyclical market. A successful construction enterprise needs to be organized to go up and down in sales to cope with cycles and other market conditions, and never feel compelled to chase inappropriate work just to maintain volume. A contractor should be profitable in good markets and bad, and any drive (desire) for size should be substituted with a drive for prosperity. The *flexible overhead* concept is a method of managing overhead that enables you to reduce it by 15%–25% in a week or less. The process allows you to do that without disrupting your organization or losing core key people. The 15%–25% range is dependent on individual circumstances and choice. Most organizations start the process small in order to be sure that it works, and then increase toward the upper limit. Flexible overhead can be introduced during a growth market or at the end of a declining market where it may actually be easier if a firm has been forced to lay off personnel (as many will before this crisis is over). Implementation requires minimal effort and the benefits are immediate, such as being able to make profit in good markets or bad and experiencing the satisfaction of never wasting hard-earned profits from good years to pay for surplus overhead you don't need in bad years.

It is natural and understandable to resist voluntarily giving up hard-earned growth. When I ask a contractor audience to select one project from a prior year that they wish they never had, most pick a job that lost money. They are wishing for smaller volume and more profit which suggests that downsizing is an appropriate choice. The marketplace is unpredictable, affected by so many variables that it is difficult to accurately forecast even a few years out or to predict the next down cycle (Figure 4.1). While a growing enterprise cannot be sure of a favorable market in the future it does need to add overhead to expand. That risk can be controlled by putting on overhead that can effortlessly be removed if the market turns down. With a portion of your overhead being adjustable, you are no longer compelled to maintain sales at any cost but can fall back if necessary and concentrate on profit.

Companies using flexible overhead manage their profit, not their volume. They can also manage sudden increases in volume of 25% or more because they are able to adjust overhead resources up or down quickly, temporarily and economically faster than the average firm can secure permanent resources. This practice is a departure from the historic norms but is clearly the profile of the successful contractor of the future.

The research has been completed, measurements made, and the reality tested. *Flexible overhead* is no-cost or low-cost insurance against profit shrinkage and losses by being able to shed overhead instantly in a market cycle downturn. It also enables you to turn off overhead costs for a few weeks or months for short-term eventualities that slow the work such as when a major project is delayed, when an exceptional number of bids are missed in a row, or when for any reason work falls off. Continually profiting in good times and bad offers a whole new appreciation for the construction business.

Figure 4.1 The marketplace is unpredictable, affected by so many variables that it is difficult to accurately forecast.

The successful contractor of the future will modify their business model to manage overhead costs in order to profit during the inevitable market downturns. In a declining market, backlog falls off and aggressive competition for the lesser amount of work drives profits down. Most firms experience a reduction in volume and profit and, despite resistance, will eventually be forced to reduce overhead personnel.

It takes some time to understand and to decide to expand overhead management, and to be prepared to introduce it. Overhead is neither bad nor good. It is neutral, a necessary cost of doing business, and as such should be as aggressively managed as production costs. However, in the construction industry, overhead costs vary dramatically from one company to another to the extent that there is no accepted standard, and in some cases, it is a loosely structured expense. Overhead is a control function, but, in some companies, it has expanded to comfort where, for example, standard office facilities and company cars can get the job done but luxury offices and cars can get it done in greater comfort. The authors are not assessing, just illustrating. Overhead should be scrupulously managed, reevaluated annually, and the determinant should be necessary versus unnecessary.

When a declining market eventually recovers, as they always have, overhead will have to be put back in place to deal with increasing sales. During a downturn, most firms will have laid off people, but often the very best employees will be retained. Most say, *"all my people are great, or I wouldn't have them,"* but the reality is that organizations mirror the population. Some managers perform good, some very good, and some great. No one lays off their best performers so the least of the good will be gone and the others will be doing fine, but as more work comes in they will need help. The choice is to put the same positions back in place or to maintain the existing core high-performance team and supplement them using flexible overhead principles. The added overhead will require no long-term commitment and can be reduced within a week. Every time work falls off, short-term or long-term contractors who have embraced flexible overhead will be glad they did. Coming out of a downturn is the easiest time to initiate *flexible overhead,* because it requires no displacement of core personnel – just the addition of non-permanent employees and resources.

As an example, let's look at one department of a self-performing general contractor doing $40 million a year. (The principles are the same for a firm doing less than $5 million or more than $100 million.) In the accounting department the CFO had four assistants with separate responsibilities, i.e. accounts payable, accounts receivable, payroll, general ledger, and so forth. Work slowed to the point that one had to be laid off. With minimal cross training, the remaining three took over. In the subsequent recovery with more work coming in, the three needed help. The choice was to reemploy the same person or skill set or to reconsider the entire accounting department process. All skilled employees spend some of their time doing non-skilled work such as copying, filing, and running errands. Instead of redividing the skilled activities among four people, a part-time clerk was hired (or you can use a temp agency) to assist the three with the non-skilled portions of their work which expands their time for the skilled activities. This new position was engaged with a written understanding that it was temporary as-needed employment and that hours worked may be reduced or increased or the position terminated with one or more days' notice. We can hear the objections already and our response is: this is being done all over the country and it works. Just try it. They didn't need a skilled bookkeeper to make copies or run errands. If growth eventually requires this position to become full-time, it may be appropriate to make it a permanent position and use part-time temporary help to continue the process. Operational planning should concentrate on needs and positions, not on employees.

Using the same guidelines, you can reengineer the workflow of every department such as marketing, administration, the shop, the yard, and most controversial of all, estimating. Everyone draws the line at estimating, protesting, *"it is far too complex, the lifeblood of the company, an extremely skilled position so temporary people can't help,"* and so forth. However, estimators spend hours a week making copies, filing, looking up products, sending out drawings and specification, chasing down vendors and subcontractors, and the list goes on. Ask any estimator if they could use a gofer, a messenger, or a runner to lessen their workload, increase their output, and improve their productivity – all of which tends to enhance their accuracy. The overall company-wide business objective should be to maintain and support the highly skilled core people in good markets and bad.

The same process is used on the production side of the business. There are as many ways to organize and manage field operations as there are companies, so no single example is illustrative. Once flexible overhead guiding principles are applied in one area of an organization, managers come to understand how they work and begin to apply them in other areas in unique ways that best suit their organization. Flexible overhead is a state of mind as much as it is a process. Managers begin to analyze workflows identifying the skilled activities required to accomplish the throughput and the unskilled activities necessary to assist, support, and facilitate the skilled core team. Almost every position in a construction organization can be augmented by having less-skilled people supporting the skilled core employees. It is economical and, once you learn how to use it, more efficient. Of greater importance, it is temporary, able to be shed in less than a week, and the employment of your core team is secure. (And they know it!)

Another benefit of embracing flexible overhead is that it enables you to profit from unexpected growth opportunities because the skill set used to reduce overhead quickly also shows you how to engage temporary resources on short notice. If the right project comes along when you are already busy, you will have the skill set and experience to temporarily expand the capacity of your core group with additional temporary support that can be disbanded when no longer needed.

One of the things common in the construction industry is to support permanent and full-time overhead for the new and larger company that we desire to be. However, the company that will be successful in the coming decades must have the ability to do less work in one year, more work in another year, and perhaps more work in the year after that. The successful company will be driven *by* the market, instead of driven by demanding a steady volume of work *from* the market.

Increased overhead equates to increased capacity which is common and appropriate during growth periods. Unfortunately, these expenses are much easier to put in place, than to get rid of. In a cyclical market, increased overhead becomes a burden and creates losses. A common reaction in a declining construction market is to search for work in unfamiliar markets, such as new geographic areas and types of projects, which may seem like the only solution to hold the gradually built organization together, but often leads to cuts in profits, and sometimes losses. A declining market obviously poses risk for a construction organization, but dealing in unfamiliar markets or new types of work simply magnifies the company's risk. Needless to say, it is dangerous to combat risk by taking on more risk.

Missing a few consecutive ideal projects, which can happen even in a growth market, or delayed project starts affects profitability. Contractors are often tempted to underprice work or look in other markets to maintain volume, rather than cut overhead. Construction firms claim they are forced into unfamiliar territory; however, the reality is that there is an alternative: reducing volume (and capacity) to adapt to market realities. This is, in effect, "cooperating" with the market rather than "fighting" the market.

A far superior and significantly less risky strategy is taking work only within the organization's proven track record. When our familiar type and location of work declines, there is far less risk reducing overhead to maintain continuous profitability. Using this approach, a contractor can manage risk to achieve success, rather than react to changes as they come along. The authors fully realize this is a huge paradigm shift from the often-heard standards: "If you're not growing, you're going backwards" or "I have got to keep my organization together for when the market comes back." The authors answer these claims by asking, "At what cost?"

Lengthy and painful down markets in the past have shed light on the reality that excess overhead costs have weakened numerous construction companies financially, which does little to prepare the company for when the market returns. The most effective practice to counter overhead challenges during down cycles is using flexible overhead, which is detailed in the following section.

The Practice of Flexible Overhead

The answer to overhead management during down cycles is a strategy the authors call "flexible overhead." Many companies already use it, although not necessarily under that name. These are firms that are capable of becoming smaller (in some cases a lot smaller) and remain profitable during periods of market decline (Figure 4.2). They obviously do not earn as much money as when they were larger, but they do not lose any either and are able to maintain their profit percentage. The reality is that even though these firms become "smaller," they are still profitable, while a significant number of their competitors are losing money.

Flexible overhead is not nearly as painful as downsizing. Under the concept of flexible overhead, contractors should engage a percentage of all overhead costs in a manner where costs can be turned off and the expense ceases in a week or less. In some cases, expenses cease in one day. A portion of overhead commitments is not taken on as a permanent expense, but as rentals, temporary personnel, interim office space, temporary employees, and so forth. The percentage range varies from 15% to 25% of total overhead, because each organization is different. Most start small, realize the advantages and increase toward the limit.

Figure 4.2 Firms adopting the flexible overhead strategy are capable of becoming smaller and remain profitable during periods of market decline.

Flexible overhead is easy to use once you try it. Learn to use temporary employees and temporary services and agencies for some portion of clerical, administrative, accounting functions, and literally every personnel function. Use short-term rentals for some portion of office and field equipment and short-term office leases, even temporary trailers, during growth stages until a new plateau of volume can be reasonably assured. Even management people can be brought on with specific company growth and performance goals associated with their continued employment. These types of services used to be available only in big cities but can now be found almost everywhere. This is obviously new and may be challenging at first for managers and team members, but it manages most of the risks associated with growth and protects the employment of existing people. This practice has been successful with established companies, as well as start-up firms, and is being embraced by a growing number of construction enterprises.

There may be minor costs associated with a flexible overhead as lease and rentals may cost more than purchased equipment, temporary employees may cost more than permanent full-time employees, and efficiency could suffer if not managed well. But the reduction and control of risk are well worth a modest additional expense. An added benefit is the motivation of existing management people who get involved and excited about this prudent, realistic, and businesslike approach to growth. Existing managers can easily discern the positive impact flexible overhead has on their job security.

Flexible overhead should be allowed to temporarily cause some cramped quarters or less comfort than, for example, private offices, but those who use it to control risk during incremental growth phases say they sleep a lot better when they get home at night. Putting on permanent overhead in a fickle market is just too dangerous. Most who have tried the flexible overhead approach have been impressed with it to the degree that they are now reluctant to add permanent overhead, and when needed they do so even slower than would normally be considered safe. They are committed to keeping some portion of their overhead flexible at all times as a hedge against a market slump, and that portion seems to grow as they realize how easy and economical it is. The modest added cost, if any, is not unlike an insurance premium for protection from a known and measurable exposure. Companies that embrace flexible overhead manage their profit and not their volume, and once they become experienced in it, it costs nothing and enhances profit.

Flexible overhead prepares a construction enterprise to do 15%–25% less volume at any given time, while at the same time preparing it to do 25% or more work and have no permanent increase in overhead risk either way. It bears repeating that an organization skilled in flexible overhead can gear resources up and down temporarily and more quickly and economically than an average construction company can secure permanent resources. This practice is a departure from the accepted norm, but it is clearly the profile of the successful contractor of the future.

Some contractors may complain this is impractical and too expensive, temporary personnel are less qualified, it is cheaper to own equipment than to rent, and so on. To these criticisms, the authors say, "that the market we deal in is cyclical." Continuous growth in sales, year after year after year, cannot last indefinitely. A focus on growth in profits, rather than sales, makes more sense. When maintaining size is the chief motivator, overhead tends to grow. However, when growth in profits is the primary focus, selection in projects shifts from a desperate need for sales to what the organization does best. Team members still have job security because flexible overhead will preserve the company and its staff. A company concentrating on profitability and not size has a different understanding of the market and their place in it. The company can still enter new markets with careful attention to their level of participation, measure of associated risk, and appreciation for the team's expectations. There is a learning curve, but a new definition of future contractor success has evolved.

This is not to say there is anything wrong with revenue growth or that it is not advantageous. There is time for revenue growth with the realization that it may be temporary or may sacrifice productivity and efficiency which usually lowers profit margins temporarily. A company should focus on growth in keeping with its vision and business plan (as discussed in the previous chapter). For example, if the business plan states that for the following year we will capture a bigger market share at the risk of sacrificing some margin and we have a financial plan to back that, then it is probably the right path for the company, particularly if they plan to recover their margins in the following year. If this is the case it is obvious they know what they are doing. In such cases, growth is justified. The authors' message is that you should not grow the business at any cost or without a detailed plan. Growth is a strategy that needs to be managed very closely with an eye on profit and an understanding of the potential for inefficiencies. Absence such detail one should focus on profit instead. Being selective in the choice of work and understanding the company's core competences can lead to organic growth without sacrificing profit. It may be slower but presents a lot less risk.

Construction companies traditionally held to the belief they had to keep together their entire organization during slow periods. Because it took so long to find and train their staff and get them to work as a team, some contractors believe it would be impossible to replace those workers when the market rebounded. During market downturns, almost all construction organizations have been in the position of eventually cutting overhead. However, the real question is whether they should have made the cuts earlier. On the other hand, during market recovery, the challenge is to determine what overhead to put back into place, in what order, and when. Should the overhead have been added in the first place? Did it enhance efficiency and profitability? Or was it just a way of maintaining a comfort zone, safety margin, or ambition?

It is preferable to think in terms of "flexible overhead" to deal with the reality of a cyclical market and to avoid the desperate need for sales to achieve a drop-dead number. To exploit a short-term opportunity, a construction company may want to consider renting and leasing equipment that goes back when the work is completed and hire temporary office and accounting staff which can be discontinued on short notice. Even if these options are more expensive compared to the alternative, the difference in cost buys flexibility and peace of mind. It is survival insurance.

When work slows down, the safe response is to avoid projects that contradict the group's goals, and work that does not fit a firm's core competencies. One option is to shrink the work area closer to home to allow thorough supervision to increase efficiency and profit. Leadership is about growing and enhancing *people*, not just sales.

Magnitude and Types of Overhead Reduction Research

The authors and their collaborators investigated overhead reduction during down cycles and conducted an industry survey to discover what overhead expenses construction firms chose to reduce during the 2008–2013 recession (Smithwick et al. 2017, 2018).[1,2] A total of 437 construction professionals responded to the survey over two months (February 2015 to March 2015). The overwhelming number of respondents included contractors and represented firms of various sizes, in terms of full-time employees (FTEs), as shown in Figure 4.3.

The first part of the survey requested the respondents to classify their company's percentage reduction for nine typical overhead categories, such as bonuses, charitable gifts, business development, and travel. The second part of the survey collected demographic information about the respondents, including annual revenue, number of full-time employees, and business sector.

The results of this study are informative, as they help your company to anticipate what cuts could be targeted in future market downturns. The analysis revealed more than nine out of ten companies cut some overhead. Almost one out of five respondents, 18%, cut between a quarter and half of their overhead, and 25% of respondents cut more than half of their overhead, as shown in Figure 4.4.

Ninety-two percent of the respondents reported that their companies reduced overhead in at least one category, by an average of about 15%. Most companies

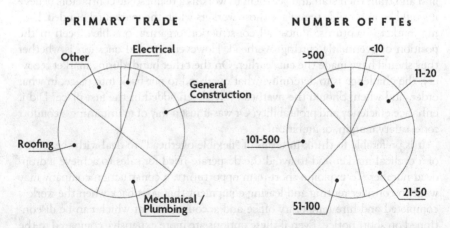

Figure 4.3 Distribution of respondents' trades and number of FTEs.

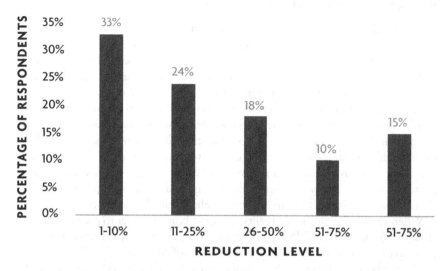

Figure 4.4 Distribution of overhead reduction levels.

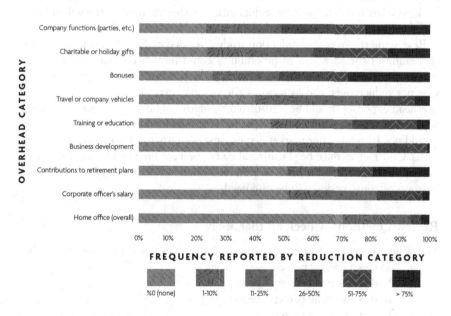

Figure 4.5 Overhead reduction by category.

reduced overhead in five different categories. Figure 4.5 shows more than three out of four, or 75% of respondents, reduced bonuses, company functions, and charitable donations. A sizable chunk, about 50%, reduced business development. Home office overhead had the fewest reductions (70% of respondents did not make a change).

Recommendations for Managing Overhead

General and administrative expenses, commonly referred to as overhead, are obviously necessary to the efficient management of an organization. For a cyclical industry such as design and construction, in which sales often fluctuate considerably, the question is how much overhead is necessary or appropriate relative to sales volume, which differs each year and sometimes each month.

A large percentage of overhead expenses is made up of salary and related costs, which are almost always considered permanent positions, necessary to the management and potential growth of the organization. With so many small to midsize closely held companies, construction contractors often run their companies as if they were a family rather than a business. This approach is noticeable when it comes to bonuses and perks, especially during good years when profits are often shared more generously than in larger or publicly traded enterprises.

When the market slows, the reality sets in: it is much easier to put on overhead than to take it off. Many closely held enterprises simply accept the resulting poor performance, and even financial losses, as a cost of doing business.

The authors recommend an alternative: a successful contractor will manage overhead through much stricter controls, and ease the pain of cutting overhead by maintaining 15%–25% of all overhead as "flexible overhead" that can be scaled back very fast to allow for volume fluctuation, rather than fill in the volume gaps with low-profit or no-profit work.

Reviewing the overhead reduction study discussed earlier, the evidence is clear that some of the sacrifices in profitability could have been avoided by cutting overhead sooner. In other words, if overhead reductions are a necessary defense, why not reduce overhead as soon as the market turns down? Every category of overhead expenses should be scrutinized regularly, and overhead costs need to be adjusted at least annually in anticipation of the following year's sales. Decreases should not be made after market downturns, but rather in anticipation of downturns. Contractors must be planning ahead to prevent financial losses and remain profitable. In the following set of chapters, financial planning practices for the business of construction will be discussed.

Review Questions. Check all that apply.

1   When a construction company has reduced sales from prior years it usually:
    a   Makes a greater percentage of profit
    b   Makes less percentage of profit
    c   Makes the same percentage as prior years
    d   Does not have a pattern

2   The construction market:
    a   Is cyclical
    b   Is not cyclical
    c   Varies slightly either way
    d   None of the above

3   Reduction of overhead for a construction organization is:
   a   Easy
   b   Not advisable
   c   Costly
   d   None of the above

4   Flexible overhead prepares a company to profitably perform _____ less work:
   a   50% or more
   b   35% or less
   c   15%–25%
   d   Any percentage

5   Flexible overhead includes:
   a   Labor costs
   b   Equipment rental costs
   c   Rearranging job responsibilities
   d   All of the above

Critical Thinking and Discussion Questions

1   Explain the construction market cycles.
2   How do you institute flexible overhead in a construction organization?
3   Explain how flexible overhead works.
4   Explain the limits in percentage size change of flexible overhead.
5   Explain the advantages of flexible overhead in a growth market.

Notes

1 Smithwick, J., Schleifer, T., Sawyer, J., and Sullivan, K. (2018). "Quantifying the Impact of the Great Recession on the AEC Industry – A Call to Reevaluate Home Office Overhead Costs," *International Journal of Construction Education and Research*, ASC, Vol 15(3), pp. 198–215.
2 Smithwick, J., Lines, B., Sawyer, J., and Sullivan, K. (2017). "An Analysis of Construction Overhead Expenses during the 'Great Recession'," *2017 Associated Schools of Construction Conference*, Seattle, WA, April 6–7, 2017.

Part II
Financial Planning

5 Using Financial Data

Following over 30 years of researching, identifying, and cataloging the causes of construction business failures and participating in the resolution of hundreds of distressed firms, the authors discovered a remarkable truth. Many construction company failures were predictable, and, in many cases, the in-house accounting staff saw it coming two or three years before it happened. A key reason some contractors experienced financial trouble was the lack of listening to the CFO or accounting staff who tried to sound the alarm before disaster struck.

Unfortunately, some CFOs are isolated and considered to lead a less important function of the business. The authors have regularly heard from contractors, *"They don't know anything about construction."* In response, the authors asked the contractors, *"Do you know anything about accounting or finance?"*

As mentioned earlier in the book, construction companies have three primary functional areas, (1) get the work, (2) do the work, and (3) account for the work; all three are critical to business success. After a contractor gets the work, they still have to do it efficiently; and if they don't accurately account for the first two functions, then they won't be in business for long. Therefore, accounting for the work is as equally important as getting the work and doing the work. However, many managers do not value the accounting and administrative functions of their business as a critical element to the firm's success.

Construction business failures don't happen overnight; they can take years. In the hundreds of failures the authors have studied, most of the CFOs warned management about financial issues, problems, or exposures before they occurred. To which management often reacted by saying, *"You don't understand"* or *"You don't know what you are talking about,"* or *"It's just that one bad job"* or *"that unreasonable owner"* or *"that unbuildable design."* These managers saw the problem as an "event," but it was a "symptom." In some cases, the CFO and accounting staff believed the excuses, but when the issues persisted, they would sound the alarm a second time and a third time, resulting in more forceful rejections by management. To provide an analogy: it is like a person in the boiler room of a ship calling the captain to tell him or her that the ship is leaking, and the captain on deck responding, *"Everything looks good from here."*

DOI: 10.1201/9781003229599-7

Communication Is Key

In more than half of the construction business failures the authors have studied or dealt with, the CFO or in-house accountant stated, "*I knew there was a problem*" or "*I knew we were in trouble.*" When the authors ask the CFO or accounting staff, "*Why didn't you tell them?*" they responded with some version of, "*I did, but they wouldn't listen.*" In one particular case, a CFO responded to the question by saying, "*The CFO that I replaced told them and got fired.*" Additionally, in many failed companies, the CFO was not on the organization's executive committee and sometimes not even included in senior management team meetings. This is not uncommon in small and midsize construction firms but was also true in some of the large company failures the authors have seen. CEOs and CFOs of construction firms seem to view things differently, often creating a communication barrier between them.

In the construction industry, the accounting function is not always given the status and significance it deserves. One reason may be due to the different personality traits of individuals who study accounting and finance to take the exceptionally challenging Certified Public Accountant (CPA) exams, in comparison to construction professionals who often spent years of laborious training on-site and rose through the ranks. The differences in experience type, training, and personality can complicate communication between accountants and construction professionals because they don't always converse the same way or use similar words; they may view the business differently and with a different order of priorities, and one personality type tends to have less patience than the other.

Research suggests there are major communication barriers between the typical personalities of CEOs and CFOs. Readers may be familiar with the standard DISC test that breaks personalities into four categories: (1) *dominant*, (2) *inspiring*, (3) *supportive*, and (4) *cautious*. The authors' observation of hundreds of CEOs indicates they are primarily *dominant* with some aspects of *inspiring*; CFOs are primarily *cautious* with some aspects of *supportive*. The differences are further complicated by some CEOs and CFOs being people-oriented, and others task-oriented. It is generally accepted that people with similar personality traits communicate better with each other than they do with people with different traits. For example, a *dominant* individual will communicate fairly well with an *inspiring*, less with a *cautious*, and even less effectively with a *supportive*.

People also differ in how they process information, speech patterns, body language, and how assertive they are (Figure 5.1). These differences can create miscommunication, misunderstanding, and even distortion of the message. *Dominant* individuals usually place a lot of emphasis on what they consider more important, while some *cautious* people place the same emphasis on both important and less important material. The *dominant* personality type can miss the seriousness of information that is presented without special emphasis placed on it.

There are also potential trust issues. It is human nature to trust people with similar personalities more than those with different personalities. Trust of a person with a different personality may not be immediate or may lack full confidence.

Figure 5.1 People differ in how they process information.

The authors' experience suggests many CEOs do not have unconditional trust in the numbers they are given, which suggests that they will make their determination as to what the numbers mean. To complicate matters, CEOs may not realize that they don't fully trust the CFO's numbers, which makes reaching common solutions more difficult.

Both communication and trust may be affected by the CEO's final say in decisions and their level of authority above the CFO. It is fair to say that CEOs are successful because of their *dominant* or *inspiring* style, and CFOs are successful because of their *cautious* or *supportive* style. However, some CEOs consider the CFO as a "service" role and expect to receive the data with relatively limited comments, intent on determining the meaning themselves. This is a persistent error in our industry, which is too widespread to be ignored. The roles and functions of the CEO and CFO must be clearly identified within each organization.

Levels of Leadership

The critical importance of the CFO in construction enterprises of all sizes should be fully understood, clearly defined, and elevated to top management if that is not already the case. All complex organizations require three equal leadership disciplines to operate profitably and manage risk. The comprehensive leadership team is made up of:

- The strategist: The big picture person, the visionary, usually the CEO.
- The operator: The manager, the person who executes the vision, usually the COO.
- The verifier: The person who measures the vision and the plan by reducing the business to financial statistics, usually the CFO.

Based on the authors' experiences, one individual can't possess all three skill sets in equal measure. In the best-case scenario, one exceptional individual may possess two. It is a mistake to expect the CEO to interpret the financial numbers, the same as it would be a mistake to expect the CFO to create the vision. The CFO is responsible for analyzing the data and producing the financial reports; therefore the CFO is usually the most qualified to interpret the meaning of the financial reports. Some CEOs argue that they themselves should interpret the financial reports, but they typically evaluate the meaning of the numbers from an entirely different perspective than the CFO. The authors are not suggesting that the CEO's interpretations will be erroneous; but rather that they are not envisioned in a similar fashion to the CFO's interpretation. The typical CEO evaluates financial reports in terms of their impact on operations; while the CFO evaluates financial reports in terms of their impact on the current and future financial health of the company. The numbers are historical facts; the financial reports are mathematical facts, and their meanings financial in nature. It is important to note that the future financial health of the company is one of the most critical issues management should be monitoring, as it indicates whether the company will remain in business.

The CFO of the successful contractor of the future will be a senior top manager involved in all strategic decisions of the organization, a member of the executive committee and of the board of advisors or directors if there is one. The authors strongly recommend boards as a successful practice for contractors of all sizes. The CFO will be the primary interpreter of the financial reports and respected by all concerned for his or her critical contribution to the company.

Construction professionals need to remember that numbers generally don't lie. Numbers tell the financial story – they don't deal in "Yes-buts" and typically cannot be explained away. Words don't change mathematics. The construction company business failures that the authors have experienced weren't caused by mistakes in accounting, but instead by mistakes in the reaction to the financial numbers. In the next section, the authors will delve into the financial numbers further using examples, and then introduce a financial performance self-evaluation method that can be used by construction enterprises.

Case Study

A very large national construction firm was in serious financial crisis to the extent that their surety became responsible for the completion of their projects spread over multiple states. During the initial investigation, the CFO was asked how the huge financial losses could not have been indicated in their accounting system. The response was that they were and for a long time. The next question was why the president of this construction enterprise was not told about the extent of the problem. The answer was:

> He was informed by the former CFO over 18 months ago but did not believe the amounts, was convinced there was more money in the uncompleted work and told the CFO if he could not account for the true profit in these projects he would find someone who could.

The CFO was fired. When the current ("replacement") CFO was asked why he could not convince the president about the depth of the losses, he replied:

> I tried and my numbers were violently disputed with veiled threats about getting it right. After seeing what happened to the last CFO who tried to explain the bad news, there was nothing left to do but hope the president was right.

One litmus test the authors use to determine if a construction entity is well managed is to find out if the CFO is a respected active member of the top management team. Recognizing "structural" financial flaws in a construction company of any size requires the active participation of a knowledgeable construction accountant within the top management of the organization; and that is doubly difficult for small organizations because they often cannot afford to have one. The needed education background, training, and experience of a construction entrepreneur or senior manager to be successful almost preclude much time spent in business courses, let alone advanced accounting or finance education. Construction is a complex, hands-on, high-risk endeavor where you have to learn to do the work right and for a profit. There is little time for the accounting and administrative activities critical to supporting the construction endeavor, with both attributes and qualifications rarely found in the same person. To ideally run a sophisticated construction operation, one needs fully qualified CEO and CFO who can work together.

Critical Need for Accounting Input: Misunderstanding Financial Information

Very few have greater respect and admiration for construction professionals than the authors, and at the same time very few have interacted with, researched, and studied as many construction professionals, projects, and companies. There is a large number of construction professionals that struggle with financial statements. The common term "to read" a financial statement is insufficient: most construction professionals can read a financial statement; fewer have an in-depth understanding of it; and even fewer can genuinely dissect and analyze a construction firm's financial statements. It is critical for a top manager to have a thorough understanding of how much they truly know about the presentation and meaning of financial data, and how much they don't know. To demonstrate that such knowledge varies among top management and contractors, a few actual examples may help.

Example One

A midsize contractor was in the habit of having his bookkeeper make checks out in advance to subcontractors and vendors so that when project payments came in, the checks could be mailed out quickly. Over time, the company grew considerably and, unknown by the contractor, was underperforming. As cash flow tightened, the checks made out in advance had to be sent out later and later and eventually most sat in the bookkeeper's desk for a month or more. Because the monthly accounting reports did not indicate the extent of the problem, the contractor was not overly concerned. What he had not realized or had forgotten was that when a check was written the accounting system listed the corresponding invoice as paid, and with so many checks sitting in a desk drawer, account payable was dramatically reduced and the accounting reports indicated profits were being earned. The contractor believed, and stated to his surety representative, that his firm was experiencing a temporary cash flow issue.

Eventually the surety sent one of the authors to investigate the situation. The contractor was asked to instruct the bookkeeper to allow us unlimited access to the books and records, which he did. The following day we were explaining to the contractor the extent of his losses for the year to date, and he showed us accounting reports that indicated that was not the case. He said that our accounts payable was incorrect and vehemently demanded to know where that number came from. We explained that the total of the checks that were not mailed amounted to hundreds of thousands of dollars, none of which was able to be covered by the minimal amount in the checking account, so the unpaid amounts had to be added back into the accounts payable which is where the number came from. After berating the bookkeeper for showing us the unmailed checks, he asked him why he did not tell him about this issue. The bookkeeper's response was that every time he handed him the monthly accounting report, he actually did tell him and even attached a note that said the accounts payable reports do not include the unmailed checks and their respective amounts.

The lesson here is that the contractor was busy managing and growing his business – as many are – and was not knowledgeable about the internal workings of his accounting systems. What he said that day was *"That's what I have book-keepers and outside accounts for,"* and he meant it sincerely. He could "read" the accounting reports which he relied on, but what he did not realize was that to truly understand the accounting reports, one needs to have some knowledge of how they are developed. He later said that when he received those notes about the checks he thought *it was a cash flow issue and did not relate it to the profit and loss* (what he meant was the income statement).

Construction is a complicated business, and this well-educated and experienced contractor had mastered it without a great deal of accounting or finance experience. This suggests that the contractor understood the construction business, but to be able to "manage" the business, he needed an accounting partner, that is, a CFO. In this case, his bookkeeper was, in effect, his CFO, but he did not think of him as a CFO, nor did he relate to the information he gave him. The financial aspects of the construction business are an integral part of the business, seldom found in the same person who directs and manages the work, and therefore must be added into top management, usually in the person of the CFO.

Example Two

A contractor in a very remote area of the country with jobs spread over three large states drew the attention of his surety after numerous complaints about partial and non-payment of subcontractors and vendors. During the investigation it was discovered that when invoices arrived at the main office, they were not recorded until approval by field personnel. They were placed in an outbox where field superintendents would pick them up on Saturdays if they were back in town from their remote projects. The superintendents would then bring the invoices to the job trailer at the project the following week to review and approve. Often, they would not approve the amount, but write on the invoice the amount that they did approve. At the end of the week, they would return them to the office if they returned to town that week and place them in an inbox.

On Monday the invoices would be circulated to the inboxes on the various project managers' desks. Project managers spent most of their time traveling to projects, but when they came into the office, they would usually mail the invoice back to the sender with a note that only so much would be approved. Some invoices resulted in further negotiation, some had other steps, but eventually each invoice was returned to the accounting department when it was finally approved and signed off on. Because the person who inputted invoices into "accounts payable" in the computerized accounting system was a part-time employee, that was done only once a week on Tuesdays. If an approved invoice was put into the accounting inbox on a Wednesday, it would be recorded as an account payable on the following Tuesday. The time an invoice arrived at the office until it was approved varied from two to six weeks, and longer if disputed. Similar to the first example, accounting reports were issued with understated accounts payables, but in this case without a notice or a note.

When this was explained to the overworked contractor, he became adamant that this made no sense, and rather than use the intercom he screamed at the CFO to come to his office. As the contractor asked the CFO about the process, he became more and more agitated, came out of his chair screaming as he reached across the desk and threw a well-aimed right hand punch at the CFO. The lesson is that anyone reading financial reports either has to have intimate knowledge about how they are compiled or ultimate trust in their CFO, who, to be effective, must be an active and respected member of the top management team.

Example Three

This contractor had an excellent relationship with the small independent bank where his accounts were deposited, and the back would regularly allow the overdraft of the checking account if there were insufficient funds. This had been going on for years without issue until the contractor fell on hard times and the amount grew large and began to be constant at which point an interest-bearing revolving loan agreement was put in place. The problem was that the contractor's CFO did not enter the revolving loan amounts which varied almost daily as a liability in the firm's accounting system. The effect and results were similar to the two examples above. The accounting reports understated liabilities which exaggerated profits and the contractor thought his only problem was cash flow.

Example Four

This problem arose from the overstatement of account receivables. There were occasions when the contractor would negotiate final payments to satisfy punch list items or owner dissatisfaction. There were also occasions when unit price projects did not reach the original contract amount. During, or at the end of each of these projects, the contract amounts should have been reduced, but the CFO left them on the books as accounts receivable. It started innocently enough when at some point the funds were believed to be due and payable, but eventually were left on the books. Over time, they were thrown into a miscellaneous category, and because the number varied over time, it did not stick out enough for the auditors to question it until it got so big and was discovered to be actually "bad debt" that had not been written off. When corrected, it caused the company to show a loss for the accounting period which shocked the contractor and the surety. He asked: "*How could this happen?*" By now we trust that the reader can answer that question as well as recommend how it could have been prevented.

We are not recommending that contractors study more accounting or finance. They are too busy for that; and even if there is time for continuing education, it may be best spent on advanced construction subjects. What we do recommend is that the totally separate and critical knowledge, experience, and discipline of accounting and finance be implanted into top management in the person of a qualified CFO, and that the CFO's input be considered in all major strategic decisions.

Construction Business Financial Reality: Financial Self-Analysis

Research indicates that more than half of construction companies do not have a formal strategic plan. For those companies that do conduct strategic planning, they spend a majority of their strategic thinking and effort looking forward and planning their organization's short-term future, and many do not utilize information from their organization's history as part of that effort. An analysis of the financial performance history of a firm can provide valuable information in determining capabilities, strengths, and weaknesses; such information should be the basis for current and future strategies. To effectively decide the future of an organization, it is important to precisely establish the organization's present status, which requires a review of where the organization has been. A comprehensive analysis of the last three- to five-year history of a company's financial performance is a good learning experience, and for some it may be a little surprising. Many construction business owners deal with the good years and bad years as they come along, but may not consider their organization's average performance over time. Many entrepreneurs provide very specific reasons or causes for their bad years, while they consider all their good years as the norm. When plans made on this basis don't work out, it is often because too much optimism is built on the assumption that good years are normal years and will repeat themselves, while bad years are an anomaly and unlikely to be repeated. This complicates measuring a company's current performance against its prior performance, a necessary element in financial self-analysis.

However, it is hard to measure "success" given all the meanings "success" can have, especially in a closely held company (Figure 5.2). It can be even harder to evaluate how the firm is performing in comparison to the competition, and to measure performance against industry norms. Many contractors are working in an information vacuum. Many complain that even in good times they are never sure their organization is performing at optimum levels. Some contractors state difficulty in defining and judging their own performance because they do not know the margins for similar contractors. With over a million separate contracting companies in the USA the industry is spread out and fragmented with so many different specialized businesses, to a point that determining norms for an industry segment is a science in itself.

Measuring a construction company's performance requires an analysis of its financial performance with particular attention focused on the organization's previous performance and changes in certain financial ratios to discover trends. An internal comparison of ratios over a five-year period of a single company provides accurate information about how that company is performing against their own historical norms. After collecting information from hundreds of firms, the authors are convinced that more is gained by studying changes in financial categories over a five-year period, than from comparisons against various compilations of industry "norms."

The authors consider financial self-analysis an essential part of any construction business planning process, and it is effective for a detailed understanding of an organization. Every contractor should stop and evaluate where they stand at least annually and carefully analyze how the organization is performing. High risks in the construction business can be reduced by recognizing the strengths and weaknesses of an organization. Financial self-analysis helps the contractor understand how to best capitalize on strengths while building defensive postures for weaknesses that may not be easily changed or improved. Financial self-analysis allows organizations to obtain an effective measure of annual performance against other years and determine if average profitability and other measures of performance are improving or declining over time.

For contractors deciding against engaging in a formal strategic business planning process for whatever reason, the authors recommend they consider conducting, at a minimum, a detailed financial self-analysis. In some cases, this activity may require input from someone outside the organization, such as the company's independent accounting firm. The process consists of a detailed financial analysis of data taken from five years of the firm's annual financial statements (prepared by the company's independent accounting firm). This detailed evaluation has provided numerous contractors peace of mind and many other

Figure 5.2 It is hard to measure success.

contractors the opportunity to make some necessary changes before it was too late. The following is the step-by-step process on how to conduct a financial self-analysis.

Managing the Financial Self-Analysis Process

The first step is collecting and compiling the required financial data listed in Table 5.1. This information is taken directly from the company's annual financial statements prepared by the firm's independent outside accountants; it is not recommended to use internally prepared statements unless all five years are internally prepared. The information is to be transcribed directly from the most recent five years of annual statement available and recorded on a spreadsheet. A three-year study is the minimum length one can learn from; five years is preferred, and any duration longer than five years does not appear to enhance the evaluation.

The ratios are calculated from the formulas in the sample financial self-analysis calculation sheet shown in Table 5.2. Both the data collection and the calculations of the ratios can be assigned to the accounting department or can be carried out by executive staff if confidentiality is a concern for any reason. It should be noted that a major concern throughout this process is transcription error; therefore, it is prudent to have a second person double-check the data. As time progresses, the financial analysis can be easily updated and used in subsequent years.

The owner(s) of the company decide who should be involved in the financial self-analysis process. Some prefer to include only owners or partners, others

Table 5.1 Key to Data Required for Financial Self-Analysis

Financial Data	Description
Year	Plan year–5; plan year–4; plan year–3; etc.
Sales	Total sales for each year
GP	Gross profit
GP/S	Gross profit divided by sales
G&A	General and administrative expenses (overhead)
G&A/S	General and administrative expenses divided by sales
NP OPS	Net profit from operations
NP/S	Net profit divided by sales
NNP	Net net profit (includes profit from other sources)
NNP/S	Net net profit divided by sales
Cash	Cash plus any securities such as CDs
AR	Accounts receivable
AP	Accounts payable
AR–AP	Accounts receivable minus accounts payable
Cash+AR–AP	Cash plus accounts receivable minus accounts payable
CIEB	Cost In excess of billings
BIEC	Billings in excess of cost
BIEC–CIEB	Billings in excess of cost minus cost in excess of billings
Debt	Total debt
Cash+(AR–AP)–Debt	Cash plus (accounts receivable minus accounts payable) minus debt

include a few top managers, and some include their entire organization's planning group. Complete financial disclosure can be avoided if ratios are only used to provide relative performance from year to year.

Exploring the source data (financial statement) will determine if any non-operational events occurred, leading to "spikes" in certain years, such as financial maneuvering for tax purposes or extraordinary bonuses that may require adjustments. Since the study is an attempt to analyze the company's organizational performance, non-operational activities should be adjusted out of the spreadsheet, as this type of information is the exclusive concern of company owners.

Calculations for the Self-Analysis Process

The numbers placed into the spreadsheet can be rounded up or down, for convenience. If a firm's annual sales are in the millions, rounding to the nearest hundred thousand is appropriate. If annual sales are in the tens of million, rounding to the nearest million is appropriate. When rounding, number four and under are rounded down to zero, while number five and over are rounded up to one. After all the information is transcribed and double-checked, the initial review can begin. The authors generally recommend a two-year and five-year average to provide short- and longer-term insight into how the organization is developing over the period. There are occasions when a three- or four-year average may be added, particularly if there was an anomaly in a given year that skews the average.

Table 5.2 Financial Self-Analysis Calculation Sheet

	A	B	C	D	E	F	G	H
1								
2		YEAR -5	YEAR -4	YEAR -3	YEAR -2	YEAR -1	5-YEAR AVERAGE	2-YEAR AVERAGE
3	Sales						SUM (B3 to F3)/5	(E3+F3)/2
4	GP						SUM (B4 to F4)/5	(E4+F4)/2
5	GP/S	B4/B3	C4/C3	D4/D3	E4/E3	F4/F3	SUM (B5 to F5)/5	(E5+F5)/2
6	G&A						SUM (B6 to F6)/5	(E6+F6)/2
7	G&A/S	B6/B3	C6/C3	D6/D3	E6/E3	F6/F3	SUM (B7 to F7)/5	(E7+F7)/2
8	NP						SUM (B8 to F8)/5	(E8+F8)/2
9	NP/S	B8/B3	C8/C3	D8/D3	E8/E3	F8/F3	SUM (B9 to F9)/5	(E9+F9)/2
10	Cash						SUM (B10 to F10)/5	(E10+F10)/2
11	AR						SUM (B11 to F11)/5	(E11+F11)/2
12	AP						SUM (B12 to F12)/5	(E12+F12)/2
13	AR-AP	B11-B12	C11-C12	D11-D12	E11-E12	F11-F12	SUM (B13 to F13)/5	(E13+F13)/2
14	Cash+AR-AP	B10+B13	C10+C13	D10+D13	E10+E13	F10+F13	SUM (B14 to F14)/5	(E14+F14)/2
15	CIEB						SUM (B15 to F15)/5	(E15+F15)/2
16	BIEC						SUM (B16 to F16)/5	(E16+F16)/2
17	BIEC-CIEB	B16-B15	C16-C15	D16-D15	E16-E15	F16-F15	SUM (B17 to F17)/5	(E17+F17)/2
18	Debt						SUM (B18 to F18)/5	(E18+F18)/2
19	Cash+(AR-AP)-D	B14-B18	C14-C18	D14-D18	E14-E18	F14-F18	SUM (B19 to F19)/5	(E19+F19)/2

A "Financial Self Analysis" spreadsheet, with the formulas built into it, is available at no cost on this book's companion website:

www.routledge.com/9781032134734

The spreadsheet is located under *Support Material* toward the bottom of the page. Note that this tool works best on MS Windows machines.

The analysis does not provide answers as much as it leads to important questions. Any interesting, unusual, or unexpected financial results should trigger the question: "*why did this occur?*" The value of the analysis is that it directs those using it to areas of performance that they can learn from. If it shows a positive result, continue it. If it is ineffective or shows a negative result, discontinue it or prevent it in the future. The two- and five-year advantages offer insight into the short-term and longer-term performance. It may take a while to discover when and why to refer to the averages. The amount and quality of the information resulting from this process is directly proportional to the level of accounting, finance, and business experience that the analyst comes to the exercise with. The following discusses a typical initial review of the data and what can be learned. Referring to Table 5.2 we will discuss what a construction professional with limited accounting and finance background might derive from the information formatted in this manner.

Sales: From these numbers you can see if the firm is growing, shrinking, or if sales are erratic from year to year. Neither is good or bad, but what is of interest is comparing the data to what was planned or preferred. Is the company achieving management's intention, and if not, why not? Perhaps the plan was poorly executed, the market changed, or any number of reasons. The point is performance of a business should be as planned and management should know why it did or did not happen. Perhaps the plan was not practical or realistic, or someone did not perform as intended. The learning opportunity is to improve the quality of the planning process and to determine how to ensure conformance to the plan in the future.

Gross Profit and Its Percent of Sales: The actual profit amounts are of less interest than the profit percent of sales, which is typically a huge performance indicator that usually results in considerable discussion. Obviously more is better, but for many contractors, how it compares to the plan is of greatest interest. If we thought in advance that a certain percentage would be achieved, it is fundamental to know why it was not. Again, either the plan was wrong or the execution was not accomplished. It should be obvious why this is so important. Continuous improvement is accomplished by studying previous performance. The percentage is expected to be fairly consistent year to year, and if not, this should be explained.

General and Administrative Expenses (Overhead): There is a cost of doing business, and to be profitable it must obviously be less than gross profit. Overhead expenses can be categorized – for control or for convenience and both are discretionary. Business operations must be controlled, but there is a cost for controls, and the necessary amount of control varies among businesspeople. Example: One firm spent tens of thousands of dollars a year on barcoding and tool administration to control tens of thousands of dollars' worth of small tools. Even if half of the small tools were lost or stolen each year, they would save money without the system. If all small tools were lost they would breakeven. Introducing this system doubled the company's annual small tool expense. We offer no comment on convenience cost; just a description. Luxury vehicles versus standard vehicles, luxury office versus standard office, and so on. Overhead expenses can be necessary or less necessary. Overhead is a cost that tends to "creep" up with success and can easily get out of control over time. Therefore, every overhead line item should be scrutinized annually before being approved for the following year. The overhead percentage will hopefully be reasonably consistent year to year, but will vary as sales fluctuate. Overhead generally remains steady, but may increase or decrease during both growth and downsizing.

Net Profit and Its Percent of Sales: The overhead discussion above becomes more poignant when discussing the net profit percentage calculation. Gross profit is earned by the firm's field operations, overhead is spent by the firm's office operations, and net profit is the result. Overhead expenses can be controversial. If gross profit suffers, for example, in a market downturn, it may become necessary to cut overhead to avoid losses. This is particularly difficult because 90% or more of the overhead costs for a typical construction enterprise are employee-related. See Chapter 4 of this book for details on "flexible overhead." Net profit percentage is anticipated to be fairly consistent, and if not, then top management should know why.

Cash and Cash Equivalents: Cash measured at the year-end financial statement varies considerably from year to year for numerous reasons and has little meaning in and of itself. When combined with other elements below, it is used in the calculation of liquidity and may be referred back to during discussions among management.

Accounts Receivable (AR), Accounts Payable (AP), and Combined: AR is the amount owed to the construction firm, primarily from project owners. AP is the amount the firm owes to others, primarily subcontractors and suppliers. These amounts vary considerably and have little relationship to sales or other categories in the analysis. The variable is the timing of when receivables are collected and payables are paid. Taken separately they have limited analytic use, but together are of high interest. AR minus AP is expected to be a positive number which simply means that the firm is owed more money than it owes to others (not including long-term debt). If the reverse is true it may indicate a cash flow problem or possibly a profit shortfall. If cash happens to be particularly high, it could mean the firm is holding its receivables and not paying down payables for some reason; experience tells us that contractors normally pay their subcontractors and suppliers timely to the extent they are able to, so if this element is negative it deserves serious investigation. The next category may shed some light on the issue.

Cash + (AP – AP): This is a measure of liquidity. For comparisons, while each element may vary on its own, taken together they should be a positive number. If the number is positive but significantly lower than prior years and sales were consistent, there may be a performance weakness. Conversely, if the number is higher than prior and subsequent years, it may indicate superior performance that year. If it is a negative number it means that the firm's cash on hand and all the money owed to it are not enough to pay what the firm owes to others, which is the classic definition of insolvency. At this point in the analysis you should be gaining serious insight into the organization's performance history.

Cost in Excess of Billings (CIEB): This is an accounting calculation of costs the company has incurred but for some reason was unable to invoice for. An example would be costs incurred performing change order work that could not be invoiced because the change order has not yet been signed or issued. It might also be for amounts of work performed or materials delivered during the last days of the month that were not anticipated when the invoice was prepared and therefore not billed during the month incurred. In all cases, CIEB is intended to represent a temporary condition after which the funds will be paid out in the future and is therefore accounted for as an asset of the firm. A problem arises when, for any reason, amounts do not become billable or are not received. It then becomes a cost with no increase in revenue and is a reduction to the reported profit for that period. The authors expect this element to be minimal compared to sales; it may vary over time, and in some firms it has been zero. CIEB is an amount that, if improperly accounted for, has been known to be a loss mistakenly represented as a positive and should always be reviewed and explained by management.

Billings in Excess of Costs (BIEC): These are amounts that have been invoiced for work that has not yet been executed, and the costs for this work have not been incurred. It is commonly referred to as "front loading" and brings money into the firm that has yet to be earned. It is also referred to as "job borrow" as it is treated similar to a loan in that it is accounted for as a liability of the firm. It enhances cash flow, but if a firm is underperforming or losing money, the funds get spent leaving a future shortfall, unless more work can be front-loaded and eventually covered by profits. It is a controversial issue in that there are advantages for cash flow, but there are also the risks of not enough future revenue to cover costs paid for in advance. Here again, a year-to-year comparison allows top management to monitor any significant variations as one of the indicators for this self-analysis.

Billings in Excess of Costs Minus Cost in Excess of Billings (BIEC – CIEB): This is not an accounting calculation, and many will say the two are unrelated. Both of these elements are adjustments made within a report developed from information collected from the construction firm's internal records. The "work in progress" report involves contract amounts, costs incurred, earnings to date, and billing to date, the details of which are beyond the scope of this chapter. However, both BIEC and CIEB are calculated in the work-in-progress schedule, and it is the authors' experience that BIEC is usually a larger amount than CIEB. If that is not the case, we generally investigate further looking for why BIEC is unexpectedly high (based on our experience).

Total Debt (Long- and Short-Term, Secured, Unsecured): Debt represents what the company owes to outside entities which are usually defined in formal load/agreements and are interest-bearing. The interest or the cost of this money is accounted for as an overhead expense or, in some firms, as "other expense." Total debt is used in the next formula for an overall liquidity calculation. As a general explanation for the next category, most of the elements listed above were described as amounts that "vary for a variety of reasons." When some of the elements move in one direction, others move in a different direction. The next category is a "catch all" developed by the authors as a method to capture all variances and what we refer to as an "interim" or "survival" liquidity test.

Cash + (AR – AP) – Total Debt: This is not an accounting calculation, and more than one accountant has stated that these numbers cannot be used in this manner. This is a non-accounting survival test developed by the authors to lead to probing questions about the financial liquidity of a construction company. In a business failure scenario, all the liquid assets of a company would be used to pay off debt (to the extent possible). Cash is obviously a liquid asset. A/R minus A/P as explained earlier should be a positive number and therefore a liquid asset, and the liability of total debt in the authors' survival test would need to be paid off by the liquid assets, or non-liquid assets will need to be liquated to cover the shortfall. Non-liquid assets are typically land, buildings, tools and equipment, leasehold

improvements, and so forth. These cannot be liquidated without partially dismantling the company. This calculation for a construction entity provides the
authors an approximation of the "degree of solvency" of the firm. One year is of
limited value, but a pattern of five years provides insight (not substantiation) into
the relative financial health of the firm, and the financial direction the firm is
heading in. The individual elements in the financial self-analysis process provide
insights and indications of where to investigate further into issues mentioned in
the commentary on each element above.

Using the Information

After the spreadsheet is compiled with the aforementioned data, a meeting should
be scheduled to analyze the results. Given that many questions will likely arise,
the team should have quick access to the company's records and key individuals
for answers.

Financial performance is the starting point. A number of other topics also
should be on the agenda and discussed in detail. These include but are not
limited to:

- The company's current market position.
- What does available work look like for the next year or two?
- How does the company compare with its competition and what are its com-
  petitive advantages and disadvantages?
- Are current operations/production optimized and to your satisfaction?
- Is the structure of the organization suitable to its current workload, and will
  it remain the same or change if you plan to grow?
- Is there a clear chain of command and do all the people in the organization
  understand the reporting chain and their responsibilities?

The questions may seem fairly simple, but it is key that top management takes the
time to review these issues simultaneously and discuss the answers to these questions in light of the five-year financial performance analysis. What happens often
in a closely held business is that the owners evaluate their business personally
and informally, and because most entrepreneurs are optimistic, it tends to shade
performance evaluation. Optimism can often cause a business owner to rate their
performance on the good years, which may not provide a realistic picture of what
the average has been or what the future may hold.

Next are a couple of examples to illustrate some of the potential benefits gained
from applying financial self-analysis.

Example 1: A construction firm using the financial self-analysis program noted
that that their overhead had increased steadily over the five years at a greater
rate than sales had increased. They determined that the increases were not a
result of any added categories, but that most line items had grown annually. They
considered inflation too, but the rate was far greater. The group reviewed every
item carefully, and reduced overhead where possible. They then decided to keep
a closer eye on monthly expenses. In one year, overhead expenses were reduced

to the five-year average. The self-analysis exposed the problem, and the solution was obvious and easy. In the second year, sales improved a little resulting in the general and administrative overhead to sales ratio decreasing even further. Non-financial managers can learn from this effort.

Example 2: In another case, a contractor discovered net profit was decreasing slightly but steadily over the prior two years, while overhead expenses were consistent, indicating a deterioration in gross profit margin. This is far more serious than the previous case because it indicates a problem in operations/production, which can be much harder to rectify than controlling general and administrative expenses. After a serious discussion with all concerned parties, they realized that a number of new field superintendents with limited experience were promoted during the company's rapid growth over the two years, and they were not producing the work as profitably as the more experienced superintendents. The principals concluded that further growth would be a mistake and could put their entire operation in jeopardy. They cut down the volume by taking no new work for six months and instead concentrated their efforts on an intensive training program for new superintendents. They decided that sales should not grow until the organization returned to the average gross profit margin they had experienced in four out of the five previous years. Their goal was achieved in nine months, and was further exceeded the following year when their gross profit percentage went up by 1.5% compared to their five-year average.

In this example, the owners of the construction company were unaware of the problem until they performed the financial self-analysis. They made the common error of thinking the slippage in profit was the price to pay for growth. The owners fell into the trap of believing it was something they called "growing pains" until they realized that their future success was in jeopardy. The financial self-analysis allowed near-immediate discovery of the problem, which helped focus the company's energy on the solution.

Final Thoughts

Standard accounting is absolutely necessary for construction companies of all sizes, as was discussed in the first part of this chapter. The financial self-analysis tool is not a substitute for proper construction accounting. The self-analysis tool is geared toward company leadership for analysis and planning purposes, effective problem discovery, and to facilitate asking probing questions in the internal analysis of a construction enterprise.

A properly carried out financial self-analysis forces a realistic evaluation of the facts related to the organization's performance. The exercise is most effective in highlighting strengths to be exploited and in bringing to the surface weaknesses to be dealt with that natural entrepreneurial optimism may cloud. It is also effective in surfacing personality issues among managers that can dramatically impact an organization's performance. These issues can be left to fester for years because in many closely held businesses there is no ready forum to address them.

As a free-standing activity, a financial self-analysis is an effective tool to maximize performance and is even more effective as a preliminary to a subsequent business planning effort. After studying and discussing the financial analysis and answering the internal and external questions, the business owners should review their personal and financial goals with each other. There is a great deal to learn about an organization from what is unquestioned, unsaid, or buried. It is also effective in uncovering unstated or unsatisfied personal or professional goals among key people, which can create stress that is better channeled into motivational forces.

In the years to come, the construction industry will continue to become more complex, amplifying risk. To make profit, contractors will need to work more efficiently; however, they may not know if this efficiency is reached or its impact on the company's financial performance without greater attention to, and respect for, the accounting functions of their businesses and the people who carry them out. In the next chapter, the authors will share methods to measure financial risk and provide tools to identify indicators of potential financial distress.

Review Questions. Check All That Apply.

1   Communication between accountants and construction professionals is complicated by differences in:
   a   Type of experience
   b   Personality
   c   Type of training
   d   All of the above

2   The critical importance of CFOs in construction enterprises should be:
   a   Elevated to top management
   b   Fully understood
   c   Clearly defined
   d   All of the above

3   High risk in the construction business can be reduced by:
   a   Reducing overhead
   b   Recognizing strengths and weaknesses
   c   Hiring consultants
   d   Reducing material costs

4   A financial analysis of _____ is recommended:
   a   Five years
   b   Three years
   c   The entire life of the company
   d   None of the above

5   The most important entry in the financial self-analysis program is:
   a   Cash
   b   Net profit
   c   Cost in excess of billing (CIEB)
   d   None of the above

Critical Thinking and Discussion Questions

1 Explain in detail the reasons that most construction CEOs don't seem to communicate well with their CFOs.

2 Discuss the reason for the misunderstanding of financial information present in the text.

3 Discuss how the realities of the industry make accurate and timely field reporting difficult.

4 Explain why it is hard to measure success given all the meanings of success in the closely held company.

5 Describe how a properly carried out financial self-analysis forces a realistic evaluation of the facts related to an organization's performance.

6 Measuring Financial Risk

The focus of this chapter is on the measurement of financial risk and performance in construction companies and to look for early warning indicators of financial distress. Spending over a decade assisting financially distressed construction companies, the authors are strong advocates of recognizing and avoiding the causes of business failure. In our experience, many contractors who thought they had a minor cash flow problem were shocked to discover the depth and seriousness of their financial difficulty when it finally came to light. With that being said, it comes as no surprise that the construction industry has the second-highest business failure rate in the country, after the restaurant business.

During the process of studying the common causes of construction business failure, the authors found that the warning signs (indicators of potential financial distress) exist in the financial records of construction companies. The authors questioned why standard financial ratios, such as gross profit and net profit, do not sufficiently provide a warning of deteriorating financial conditions or increasing financial risk early enough to prevent most failures.

Profit alone is not an appropriate measure of success in the construction business. Profit is a necessary ingredient for success, but it is not a measure of success. For example, a company with sales of $10 million annually, growing to $20 million in a year, without a corresponding increase in equity, may dramatically increase financial risk without realizing it, even if it maintains profitability. Therefore, to understand financial risk warning signs, it is important to first understand the financial performance measures within the construction industry.

Financial data that will be discussed in this chapter should be taken from financial statements prepared by an independent outside accountant or accounting firm. These statements can be reviewed, audited, or otherwise qualified. What matters is that statements be consistent. Data from internally prepared financial statements are not recommended.

This chapter focuses on the discovery of early warning indicators of deterioration of a construction company's financial performance well before normal financial reporting and standard financial ratios would display them. This is important because early detection of financial weaknesses would allow management time for appropriate defensive action to be taken and prevent business failures. This chapter provides construction company executives, entrepreneurs, and credit grantors

DOI: 10.1201/9781003229599-8

with tested guidelines and a financial risk indicator to reduce business risks that previously have been accepted as inherent or unavoidable in the construction business. The primary purpose is to introduce a new method of measuring construction company financial risk and performance in order to predict financial distress or failure earlier than current financial evaluation procedures.

Traditional Financial Performance Measures

A construction company may evaluate its financial performance using various standardized financial ratios. This includes examining the organization's gross profit, net profit, equity, and capital structure. However, these standard ratios alone do not adequately reveal the performance of the organization.

Gross profit is the amount earned in the production of the work before general and administrative costs are deducted. Similarly, net profit is the amount made after subtracting all general and administrative costs from the gross profit. Both gross profit and net profit do not accurately report an organization's financial performance, as there is no standard for measuring either of these measures consistently in the construction industry. There is no agreement regarding which costs are to be charged directly to projects and which are general and administrative overhead costs. As a result, comparisons of gross profits and net profits cannot be used as accurate measurements of a construction company's financial performance.

Analysis of financial statement data is conducted primarily to measure financial performance. However, there are conflicting views whether financial performance should be measured by profit or by an increase in the value of the firm. The ultimate measure of performance is not what is earned but how the earnings are valued by the investors. A firm's increase in value, not just its profits, is the true measure of a company's performance. Therefore, it is necessary to measure the risk that the firm's value is put at in the pursuit of profit in order to measure its overall financial performance.

Equity, which is the company's assets after subtracting all associated debts, is a significant measurement anchoring the capital structure that underpins the company and supports credit facilities that provide cash flow during peak needs. Capital structure is the organization's unique combination of debt and equity used to finance its overall operations and growth. Equity alone is not an absolute indicator of an organization's financial performance. Moreover, the capital structures of construction companies differ significantly by type of work performed. Highway contractors who own their equipment have different capital structures than a building contractor who subcontracts a great deal of their work. Therefore, equity and capital structure are also insufficient measures of a construction company's financial performance.

Corporate decisions and events that are common causes of a construction company's financial distress or failure typically precede the actual failure by two to three years. However, modest changes in financial statement data and small changes in standard financial ratios do not appropriately announce deterioration in financial performance or possible increases in financial risk. The combination of the fast-paced construction business and the lengthy timeline of major projects makes early detection of financial weaknesses difficult. This is especially true in a growing business, as growth tends to cover up poor performance. When the business is growing, a good rule of thumb is to allow current liabilities to grow only as fast as sales, so that term loan repayment prospects are improved.

The R-Score Formula

Accounting tools used to interpret a construction company's financial statement do not always project the firm's actual performance because they often concentrate on only one financial area without linkage to others. The authors realized measuring financial performance and risk of a construction organization requires combining standard financial ratios in a holistic manner that provides insight into interpretation of its financial activities and financial strengths and weaknesses. This led the authors to develop the R-score formula, which:

* Measures an organization's current financial performance;
* Compares an organization's current performance to past performance; and
* Identifies an organization's current financial risks, strengths, and weaknesses.

As noted previously, changes in one financial ratio inevitably change other ratios. The interdependence of various ratios can be traced and used to develop a financial measurement formula. To evaluate a closely held construction company several variables must be considered. The problems of different bookkeeping methods and treatment of work in progress suggest that gross ratios rather than ratios found internally should be used. The significant standard financial ratio categories are turnover, profit, and debt ratios. Ratios are selected from these categories that are least affected by internal bookkeeping methods. These are explained below as the three key elements of the R-score formula:

1   **Sales-to-Total Assets**: This is known as the turnover ratio and is a measure of operational efficiency. The higher the ratio, the more efficient the utilization of assets. The ratio is a composite of receivable management, inventory management, fixed asset management, and liquidity management. The ratio is the relative efficiency with which the construction firm uses its resources to generate output.
2   **Net Profit-to-Sales**: This is known as the net profit margin and is a measure of operating efficiency after all costs and expenses have been accounted for. While the external marketplace affects the sales-to-assets and net profit margins, they largely capture an organization's internal management efficiency.
3   **Total Liabilities-to-Equity**: This is known as the debt ratio and it tests long-term liquidity. The ratio is like debt-to-equity; however, total liabilities (all debt) are used instead of long-term bank debt because the latter can easily be reduced, temporarily, particularly at year-end, by extending accounts payable

or substituting short-term borrowing. The ratio is applied here as a measure of the firm's ability to sustain itself over the long term. This is a broad ratio ignoring most internal manipulations or differences in bookkeeping methods because it captures all liabilities and equity not easily manipulated.

The three key elements shown were combined to develop the risk factor score (R-score formula), an effective and comprehensive financial risk management formula. The concepts underlying the interrelationships between the elements of the ratios and the ratios themselves are detailed in the following paragraphs.

Development of the R-Score Formula

Within the standard profitability ratios, the term "profit" seems to imply a firm can use the funds generated by profits for whatever purpose it chooses. However, this is not the case in reality, because it takes money to run the business. In other terms, the business captures and uses funds to operate; without these funds it cannot operate. The amount of money required to run a business varies by company and industry. For a closely held construction company, it can cease operations very quickly if it runs out of cash and credit. The portion that must remain in the business, to maintain the company's assets to liability balance, will vary between individual firms but is approximated by subtracting from net profit margin the proportion that liabilities represent in the liabilities to assets ratio. Simply put, if a firm's liabilities to assets ratio is 1 to 3, one-third of the profits must remain in the operation to support ongoing business. If more than that is taken out of the company or used for other than normal business activity, other funds will have to replace those taken. For the closely held company, that inevitably means borrowing. Borrowing increases liabilities with a resultant change in the liabilities to assets ratio. The amount of investment in the company held by outsiders (e.g., banks) goes up, while the investment of the owners of the company goes down. The company is at greater financial risk because it is less self-sustaining. There is a difference between net profit and available net profit.

Funds generated from profits are not entirely available for other uses because a portion must be dedicated to the following year's operations, replacement of assets, and productivity improvements to remain competitive in an ever-changing industry. Some of the profits earned also contribute to the costs of timing of receipts and retirement of liabilities. Each company has its own liabilities to assets ratio that does not necessarily relate to similar firms or industry standards. The balance of *liabilities to assets* is a measure of long-term liquidity of the firm, and a certain portion of profits is needed to replenish assets and retire liabilities in the near term to maintain the balance. The replenishment will vary depending on liabilities to asset ratios unique to each company, which is simply the financial makeup of the company or its financial foundation. This concept can be referred to as "available profit (AP)." The amount of funds generated by profits, remaining more or less permanently within the financial structure of the company, can be measured to determine the AP.

The authors define AP as the portion of profits that can be taken out of the company or applied to business expansions without materially affecting the financial foundation of the firm in its existing operations. The measurement of AP is accomplished by reducing the total profits by the proportional amount of liabilities in the liability to asset ratio or the total liabilities divided by total liabilities plus assets.

The formula for AP is:

$$AP = \left(1 - NP/S - \left(10 \left(TL/(TL + TA)\right) \times (NP/S)\right)\right)$$

where NP/S is net profit/sales, TL is total liabilities, and TA is total assets.

To address the negative profit (loss) of companies, the one minus net profit to sales, in the equation, creates a positive value in this element of the formula. Sales to total assets and total liabilities to assets are generally a number greater than one. Therefore, the decimal place for the profit element is moved one place to the right to create an appropriate relationship with other elements in the formula.

The measurement of financial risk includes the earning power of a company, where the earning power ratio is the profit margin times sales to asset ratio. However, "real earning power (REP)" of the company can be determined by multiplying the AP by the asset turnover ratio of sales to total assets (S/TA). The formula for REP is:

$$REP = (AP) (S/TA)$$

where AP is available profit and S/TA is sales/total assets.

To determine the overall financial risk of a company, its debt structure must be considered. Combining the firm's debt ratio of total liabilities to equity with REP, also referred to as the financial "risk factor," represents the financial risk of the company at the present time. The financial well-being of the company or risk is determined by using the company's own financial performance, turnover rate, and debt structure.

The R-score formula is:

$$R = (\textbf{Available Profit}) \times (\textbf{Real Earning Power}) \times (\textbf{Debt Structure})$$

And can be used as:

$$R = \left(1 - NP/S - \left(10 \left(TL/(TL + TA)\right) (NP/S)\right)\right) (S/TA) (TL/E)$$

where R is risk factor and TL/E is total liability/equity.

Using the R-score formula, the authors have produced a new financial performance evaluation method that takes into consideration the performance, capital requirements, and total debt of the individual company.

Classification of R-Score

The R-score formula was tested on 46 failed construction organizations for the three years prior to failure and for 53 successful construction organizations for the

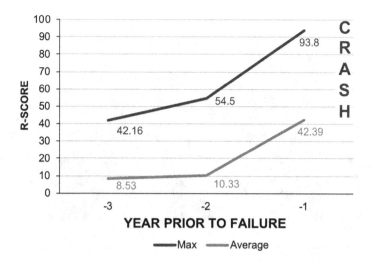

Figure 6.1 R-score of failed companies.

corresponding three-year period. The tested companies included multiple types of general and specialty contractors and the high and low scores varied (Figure 6.1).

R-scores of the 46 failed companies were positive numbers from 0.48 to 93.8. In 76% of the failed companies, the R-score was higher the year prior to failure than in either two or three years prior to failure. In the third year prior to failure, the scores ranged from a low of 0.66 to a high of 42.16, with an average score of 8.53. In the second year prior to failure, the R-scores ranged from a low of 0.84 to a high of 54.5 and an average of 10.33. In the year immediately prior to failure, the scores ranged from a low of 0.48 to a high of 93.8 and an average of 42.39.

In contrast, R-scores of successful construction companies were much lower. They ranged from 0.01 to 24.49 with an average of 4.37 in year three, 4.41 in year two and 4.18 in year one, and an overall average of 4.32 for all three years. The differences in scores between years are not significant.

The risk factor scores for companies known to have failed were consistently higher than the scores for the non-failed companies. In the year prior to failure, the average R-score for the failed companies (42.39) is ten times higher than that of the successful companies (4.32). In the year prior to failure, only 4 of the 46 (9%) failed companies tested scored lower than 4.0. Of the 53 successful companies tested for three years, only 6 (11%) had a risk factor score higher than 7.

The results of the test suggest that R-scores can serve as a measure of financial distress, as follows:

- Scores below 5 indicate low risk of failure;
- Scores between 5 and 7 indicate moderate risk;

Figure 6.2 R-scores can serve as a measure of financial distress.

- Scores between 7 and 9 indicate that financial resources are stretched compared with the company's current operating efficiency, and the firm is at high risk;
- Scores above 9 indicate very high risk. After collecting more data on 136 companies, only about 6% with a score higher than 9 did not fail (Figure 6.2).

If the R-score is very high, then the company is highly leveraged. An over-leveraged company cannot deal with a slow year or disruptions in receivables. The company may be profitable, but its financial condition is precarious. Operating profitably "on the edge" makes little sense for closely held companies for many reasons, one of which is that the principals are usually personally liable. There is too much at stake to allow the continuation of high financial risk after discovery.

The raw scores of the risk factor are significant in themselves and can be used in the current year for evaluation purposes. In addition to the actual score itself, the R-score trend over several years is also important. The trend is a valuable indicator of the financial well-being or distress of a construction company and whether its financial risk is improving or deteriorating. A multi-year trend analysis strengthens the prediction or projections concerning the relative financial strength or risks that can be known about a closely held construction company.

The failed companies tested had an average increase of 21% in their R-scores from year three to year two prior to failure. They had an average R-score increase of almost 400% from year two to year one prior to failure.

Contractors should determine their R-score for the past three, four, or five years to see how they are trending. A calculation of the previous year's score will establish whether the company's financial risk is improving, constant, or deteriorating. If, for instance, a company had an R-factor score of 6.5, and the prior year's score was 7 or 8, the financial risk is now moderate, but the trend is improving. However, if the same company had the same score of 6.5, but the prior year's score was 4 or 5, the company still has a moderate financial risk but worsening situation. A look at the score three years back provides a longer trend and can give the company's management an added sense of where the company is headed.

An "R-Score Calculator" spreadsheet, with the formulas built into it, is available at no cost on this book's companion website: www.routledge.com/9781032134734

The spreadsheet is located under *Support Material* toward the bottom of the page. Note that this tool works best on MS Windows machines.

Why the Formula Works

The R-factor formula answers three questions:

1   Is the construction company's performance adequate considering its capital structure?
2   Is the company's earning power providing enough funds to maintain its assets to liabilities balance?
3   Is there adequate equity in the company's capital/debt structure to deliver the capital or credit necessary to underwrite operations and ensure against unforeseen losses?

The Bottom Line

The R-score formula provides an easy-to-use tool for measuring financial risk and financial strength for a closely held construction company. It allows for a multi-year evaluation providing a quick and accurate historical trend of financial performance and financial risk. The R-score formula can be used as an internal self-evaluation tool for closely held construction companies or an external analysis tool for credit grantors.

The R-score will help professionals in the construction industry better evaluate the financial risk of closely held construction companies. Several financial ratios are combined holistically to provide insight into the construction operation's financial activities. This helps contractors better understand the impact of operational changes on the financial health of their companies.

A reduction in the extremely high failure rate in the US construction industry will ultimately impact the economy, given that construction is one of the largest industries in the US, responsible for a significant portion of the GDP, and provides more than 11 million jobs. It is imperative that contractors have control of their organization's financial risk.

In the following chapter, the authors will discuss liquidity management techniques to support a construction business's financial planning and management.

Review Questions. Check All That Apply.

1   What is the ultimate measure of a company's performance?
     a   How earnings are valued by invertors
     b   How much is earned net profit
     c   Gross profit
     d   Reduction in debt

2   How do the authors define "available profit":
     a   Gross profit before overhead
     b   Net profit after taxes and interest
     c   Profit taken out of the company without effect
     d   None of the above

3   The R-score formula determines the:
     a   Economic strength of a construction company
     b   Reputation of a construction company
     c   Overall financial risk of a construction company
     d   Return on investment of a construction company

4   The turnover ratio is:
     a   Sales to total assets
     b   Net profit to sales
     c   Sales to total equity
     d   None of the above

5   An R-score is strong when it is:
     a   Remaining steady
     b   Reducing over time
     c   Decreasing over time
     d   None of the above

Critical Thinking and Discussion Questions

1   Discuss the concept of the ultimate measure of a construction company's performance.
2   Explain the concept of "available profit."
3   How does the R-score formula differ from the traditional financial performance measures?
4   Explain why the s-Score trend over time is important.
5   Discuss the author's explanation of why the R-score works.

7 Financial Distress –
Turnaround Management

The US economy has always been and continues to be cyclical, therefore the construction market as an integral part of the economy will continue to be cyclical. Construction is a lag industry, which means it follows the ups and downs of the US economy after a typical lag of 12–18 months. When the construction market is growing, the industry prospers; however, when the market declines, the industry suffers reductions in profits and exposure to loss. The reason is that growth requires additional overhead expenses in the form of personnel and resources which are usually brought on as permanent fixed costs. When the market declines, it is extremely difficult to reduce fixed overhead because the company cannot eliminate half a person or half a truck. As discussed in this book, the construction business model of permanent fixed overhead is one cause of the industry's boom and bust cycle. This raises the following question: If a firm earns a certain percentage of profit putting X amount of work in place in a year, shouldn't it earn about the same percentage of profit the following year when it puts in place, perhaps 15% less work? Most construction companies don't.

Profitability. Whenever there is less work, margins suffer because the selling price of construction goes down as the industry shifts into a *"buyer's market."* Losses become more common, and some businesses eventually fail. No type of construction was spared in prior downturns because contractors with limited experience in other types of projects or in different geographic areas went after work they were not familiar with. Many were successful in capturing that work because they were aided in their pricing by a lack of experience or limited local knowledge. When the market shrinks, competition increases and the selling price deteriorates for every type of work as many construction companies fight for work of any type, anywhere, and almost at any price.

The *"Compression"* Dynamic. The authors define compression as the domino effect of the downward pressure applied to construction pricing by large contractors during market contractions (downturns). What often happens in a downturn is as follows: larger construction companies, desperate to keep their people busy, aggressively go after smaller projects than they normally would. This takes work away from other firms, forcing them to go after smaller projects than they normally would too, through midsize firms and finally down to small firms. With their greater capitalization, big contractors are able to use aggressive pricing to take

DOI: 10.1201/9781003229599-9

smaller jobs and/or expand geographically by moving into markets they do not normally compete in. The lowest level in this avalanche, the smallest contractors, have nowhere to turn to and some are driven out of business.

Defense Strategy. Contractors need to understand this dynamic which happens every downturn and should therefore be expected. Compression highlights the reality that cutting the cost of doing business and downsizing are proven defense strategies during a declining market. Cutting overhead proportionally to the amount of the market decline enables you to produce a lesser amount of work at a profit. If there is less work available and a firm attempts to maintain its sales volume, it will have to take the work off its competitors who are trying to do the same thing. Compression causes industry-wide losses which is partly the reason that construction is the industry with the second highest business failure rate in the country – a failure rate that always accelerates during declining market cycles. This has been happening for decades and has not been recognized because it was an unidentified dynamic for many contractors. Now that we know about it, we can deal with it.

The most popular choice in a declining market – chasing work while profits are spiraling down – exposes an organization to serious risks. The difference is that in a growth market there are more profits from the company's other projects to help cover the loss from the occasional bad job. In a declining market, diminished profits may not be enough to cover a losing project, and a losing project is an increased possibility because of the aggressive pricing trying to capture the much-needed work. Even during a growing market, construction is a low-profit business with the ever-present possibility of a losing project in both good and bad markets. This exposure is compounded when, motivated by the need for work, a firm decides to go after work in unfamiliar territory and/or pursue work of a slightly different type or size than the organization is experienced with, which increases project risk exponentially.

Overhead Management

One of the reasons the failure rate goes up in a declining market is that many construction enterprises wait too long before reducing overhead. This weakens their financial condition, which makes them vulnerable if the downturn drags on, and hinders their financial ability to fully participate when the market rebounds. Many have said, *"If I am forced to reduce overhead at some point(s) during a downturn I might as well have cut it sooner."* It is disappointing to have spent a lot of resources holding onto people, only to have to let them go anyway. Many have tried to keep people busy by traveling far and wide, taking work outside the organization's experience and core capabilities, but that often magnified underperformance and, in too many cases, resulted in losses. During many years of turnaround work we have often heard: *"What choice did I have?"* or *"I had to keep my people busy,"* or *"No one can make money during a downturn."* For those determined to maintain sales or to reduce sales as little as possible in a declining market, these statements are true, and we fully understand and admire the determination to retain not just some, but all, employees. This said, can they afford the cost? When maintaining high sales while profits are spiraling down, there is an exponentially increasing risk of business failure, at which point the company may lose all employees, not just some.

It is human nature to resist giving up hard-earned growth gained at great effort. The problem is the risk and cost, and it does not pass the logic test. Consider the reality: Why can't a construction enterprise increase their sales by say 10% or 20% any time they care to? Because you can only grow as fast as the market grows. If the market is not expanding at 10% or 20%, that amount of growth would be extremely difficult and expensive to attempt. They would have to drop their prices, find additional resources, secure them, and pay for them. The opposite is the case in a declining market. Severe adjustments in pricing will be forced into the marketplace and idled excess people and resources will need to be paid for or discontinued. Everyone seems to agree that growth requires an increase in overhead, but they do not seem to agree that the reciprocal is true. The reciprocal is that negative growth (a reduction in sales) requires a reduction in overhead.

Construction Market Realities

A review of the research information, data, and facts keeps coming back to the same conclusions. The US economy has been cyclical for as long as anyone can remember. The US construction market follows and lags the economy; it has been cyclical since at least World War II, and continues to be cyclical, with periods of growth followed by periods of declining markets. Conclusions also include:

- A growing construction market reduces competition;
  - A declining market increases competition
- A growing market fosters increases in profits;
  - A declining market causes decreases in profits
- In a growing market, human and other resources become scarce;
  - In a declining market, underutilized human and other resources have a continuing cost
- A consistent growing market becomes a *"sellers' market"* and the selling price of construction goes up.
  - A declining market quickly becomes a *"buyers' market"* and the selling price of construction goes down.

These truths demonstrate that the current prevalent construction business model of permanent fixed overhead may be inappropriate and even illogical. It does not, cannot, and will not support consistent profitability in a cyclical market. There is a solution, *Flexible Overhead*, which was covered earlier.

Construction enterprises generally make more money in good markets than they make or lose in downturns, and because the length of growth periods has been longer than the downturns, most come out ahead. This has been going on for so long that it is generally accepted as a cost of doing business. However, that "cost of doing business" includes the risk of business failure, as demonstrated by our industry having the second highest failure rate in the nation. The business model must be modified to a process that allows construction organizations to prosper in both good markets and bad markets. It makes no sense that a growing company that has been profitable, for example, at eight, then nine, then $10 million in sales would automatically lose money if it is forced back to nine or eight million dollars by a decline in the market. Quite the opposite may be true: if the company reduces its overhead, focuses on its core strengths, not pursuing some risky and potential losing projects, that may actually help its profitability so it can come out strong and ready to participate once the market rebounds.

Challenges during Growth Markets

When the construction market finally turns up (begins to recover) after the typically lag period of 12–18 months, the margins or profits usually do not recover for at least another 12–24 months. In some cases, following a long downturn, the recovery may be even longer. In general, profit margins in the construction industry do not recover until the "construction put in place," measured in dollars, returns to at least 90% of the amount it was before the downturn. During recovery, most, if not all, construction enterprises intend to grow back to at least the size they were before the downturn. They will then face the risks associated with growth and particularly risks associated with the rate of growth. As the market improves, one would expect a seller's market to develop; however, that often does not happen until the market returns to within 90% of the prior market. Contractors, generally considered optimists, appear to be pessimistic when forecasting a growing market following a downturn because they continue aggressive pricing until the pre-downturn volumes are in sight.

An unfortunate number of construction companies continue to experience adverse effects during a good market. Construction companies of every size who take on more work than they can efficiently handle generally experience financial distress or worse. This is significant because even the largest, oldest, and most sophisticated construction companies in the world are not immune. Simply said, "If they can get into trouble, anyone can."

Prior to the 2008–2012 market, downturn most contractors had been growing rapidly over the previous years. Contractors that did not enjoy the rebound in margins during the good market were typically trying to capture more than their share of the market, growing beyond their organization's capability. This led to the recognition that it can be difficult for a construction enterprise to project how much work they can effectively perform and finance during growth markets.

Rapid growth puts a strain on a company's people and systems because sustained growth does not allow for reasonable training periods. Of even greater concern, continued growth does not give an organization a chance to test new people or systems before these replacements are integrated. If performance deteriorates, its discovery occurs after the addition of volume and people, so corrective measures are more difficult with existing people and systems stretched and with overworked managers coping with the largest volume than the enterprise has ever handled. Some companies are unable to recover from this scenario. Too many organizations grow so rapidly they can't measure performance until it is too late.

Every organization has a growth limit. Even profitable work puts a strain on cash flow as well as on human resources, and few closely held construction companies can gear up quickly or solidly enough to maintain profit margins during rapid growth periods. In addition to considering flexible overhead, which was discussed previously, the authors identify other factors that construction companies should consider to improve their organization's performance during growth markets. These include cash flow and financial constraints, as discussed next.

Growth consumes cash in upfront investments because construction enterprises put the work in place and wait for their money. It can take weeks and often longer for the owner to approve and complete payments. If a company is continuously putting work in place in each subsequent accounting period, they can eventually run out of cash and credit. It is critical (and difficult) for contractors to accurately project how much they can effectively perform and finance. A contractor must consider, carefully, how additional work will impact the organization and approach cautiously. Discussed elsewhere, an effective way to measure the extent of financial risk is to calculate the organization's R-score. If the trend is upward, management should carefully examine the company's exposure.

Managing a closely held construction company is like driving a truck up a hill. The steeper the grade, the more strain on the truck and its engine, suspension, and drive train. In the case of a company, the strain is inflicted on the employees, systems, and finances. A truck, starting up a hill from a level roadway, will find it easier compared to starting on a grade and increasing the slope. Additionally, separate short climbs are much easier on the truck, than sustaining continuous uphill progress. When trucks attempt a steep and long hill, they are laboring to gain forward progress, and in some cases they actually stop.

When a construction organization embarks on a steep climb, with a growth rate exceeding 15% yearly, there will be strain on its resources. During periods of continuous growth, the strain is sustained and magnified, sometimes to a breaking point. Mild and inconsistent strain on company resources may increase efficiency and profitability. However, subjecting resources to severe or continuous strain encourages inefficiency, deterioration, and potential failure. There are fundamental financial constraints to profit and growth. The management of growth requires careful balancing of sales objectives with the firm's operating efficiency and financial resources. The trick is to determine what sales growth rate is consistent with the realities of the organization and marketplace. Companies have limits in abilities, capital, and available resources.

A construction firm may be altered considerably during periods of rapid growth as it becomes a new untested organization right at a time when it has a lot more work to produce. If an organization is to grow, its management must also grow. Moreover, this growth must be qualitative, not just quantitative. Qualitative organizational growth takes time and needs to occur prior to sales growth. It takes more time to grow management than it takes to capture more work, and most companies decide to grow management only after additional work is on hand.

Incremental growth, instead of sustained growth, may seem unnecessary or even unnatural. However, this form of progression is the best approach to control the inherent risk in growth beyond 15%. With a series of growth increments, you can test, then grow again and test again. With this process, an organization can reevaluate and recover after a bad test before resuming growth. This is a prudent risk-control method. In sustained growth, a company grows beyond its people and systems so often that they never really have the same organization long enough to truly test it. These organizations are at constant risk with an ever changing expanded team.

Indicators of Financial Distress

When a company has suffered poor performance, it experiences decreased profitability and equity. This can lead to increased borrowing to the extent that creditworthiness is threatened. Therefore, rushing back into a good market without a fiscal check-up can be a colossal mistake. An organization needs proper posture to grow again after a stressful period because growth can deplete cash and can strain a company to its breaking point as discussed earlier.

Most construction companies that get into serious cash flow problems do not see it coming. Financial stress needs to be addressed quickly both during downturns and growth periods. Any drop in performance erodes financial strength and increases the likelihood of financial distress, so a simple check-up test may be in order.

There are a number of indicators that can warn of financial distress within a construction organization. The following are ten common financial distress warning signs; a company enduring a combination of three or more items listed below may be experiencing, or on the road to experiencing, financial distress (Figure 7.1).

1 *Disproportionate Increase in Overhead*: Companies that are increasing in overhead often claim, "It's because we're growing." However, a problem arises, for example, when a contractor performed $20 million of work the previous year with a certain overhead and this year overhead is up 30%, but volume is only up 20%. This scenario can even be worse when downsizing, for example, if a company's sales are down 20% with little or no change in overhead. Overhead changes need to be proportionate to work.

2 *Increase in Turnover in Personnel*: Unexplained departures of key people "in the know" can be an indicator of distress when combined with some of the other indicators on this list. When senior staff members leave the organization, the contractor loses valuable institutional knowledge. It can be a sign that those leaving the company think they are on a "sinking ship."

3 *Late Accounting Information*: When the accounting information does not come out on time, it may be because the numbers are not adding up, and the preparers are looking for more data or "massaging" the data. There may be a

Figure 7.1 The ten common financial distress warning signs.

legitimate reason for the delays, but there is also a risk that something else may be happening.

4 **Late Project Financial Information**: Some of the information we get in our industry may be delayed in project timelines; however, if the delay is later than usual, it is worth further investigation. And when the information is received for an active project, any major swings in job profit, whether positive or negative, demand immediate explanation.

5 **Unexpected Borrowing**: Whether for working capital, equipment, or expansion, all borrowing activity should be foreseeable and planned for. Unexpected borrowing is often an indicator of financial distress. Why the sudden need for cash?.

6 **Increase in Internal Disputes**: Underperformance breeds discontentment. Internal disputes may be inevitable; however, they can impact employee morale and productivity. If disputes are increasing, something may be wrong.

7 **Decrease in the Quality of the Work**: During financial stress, the dedication of employees to quality work output suffers. Cutting corners on quality may be a sign of desperation.

8 **Too Many Excuses**: If more things are going wrong and fewer people can provide the reasons, financial performance is probably slipping too.

9 **Departures in the Accounting Staff**: Of the hundreds of companies in jeopardy studied by the authors, more than 40% experienced the CFO or senior staff in the accounting department leave within one to three years prior to failure. It is fair to say that if the CFO or senior accounting staff leaves unexpectedly, it is likely that there is some form of financial difficulty within the company.

10 **Inadequate Time to Perform Key Functions Well**: The tasks in a company generally follow a timeline for completion. If a company's internal performance functions such as administration and management begin to fall behind, and tasks are delayed or incomplete, it should raise a red flag. This does not mean that working six days or long hours are unusual. However, if there is limited time dedicated to cover baseline activities due to more time spent with attorneys, outside accountants, or consultants, it may be an indicator of trouble.

Financial Distress Performance Metrics

On a more quantitative level, there are many ways of measuring the financial performance of the construction business. However, during times of critical financial distress, the authors emphasize two key ratios to measure a company's financial performance: debt to equity and total liabilities to total-assets. The following is written for construction executives, not for accountants who can provide the reader with a more detailed and technical explanation.

Debt to Equity: This financial ratio should never be overlooked. If the debt to equity ratio is deteriorating, financial performance is probably slipping. This ratio should be tracked at least annually. If for two or more years it is not constant or improving, something needs to change. The ratio will vary between companies, but generally debt should not be increasing as a percentage of equity over time. If debt is increasing faster than equity it should raise the question: What is the extra borrowed cash being used for? One possible reason is to fund losses. Continuing profitable performance would leave the ratio steady or improved. Improvement is measured by a decrease in debt relative to equity. If the ratio is not steady or improving, management needs to know why. The authors have experience with failed construction firms where this ratio had been deteriorating steadily for a number of years and no one seemed to notice or those who noticed did not understand the implications. The average ratio during successful years is a benchmark to be measured against at least annually.

This ratio can be problematic as both debt and equity can be manipulated and the timing of the amounts can vary. Therefore, the authors use a second ratio as well, where "total liabilities" describe everything that is owed (in other words, total debt), and "total assets" describe everything that is owned.

Total Liabilities to Total Assets: Similar to the above, this ratio varies from firm to firm, but the average ratio during profitable years is a benchmark expected to be steady or improving from year to year, showing increased total assets relative to total liabilities. The business objective is to increase assets over time, but if the liabilities increase faster, the financial condition is weakening, and underperformance is indicated. Similar to debt to equity, the total of all liabilities is payable in some fashion and in this measurement is thought of as "debt." Liabilities should not be growing faster than assets. This is not suggesting that debt is negative or needs to be paid down. Debt is a functional part of the capital structure of a business. What we are saying is that if total liabilities are increasing at a greater rate than assets, the company is falling further into debt. Unless there is an explanation for why this is happening, the debt may be being used to fund losses.

Value versus Earnings

Financial performance is not measured strictly by what is earned, but by the value of earnings seen by the investors. Every firm's broad financial goal should be to maximize the value of the business. A firm's increase in value, not just in profit, is the true measure of the company's financial performance. Therefore, to measure a firm's overall financial performance, it is necessary to measure the risk to its value, related to the firm's pursuit of profit.

There are serious drawbacks to profit maximization as the primary goal of a firm because a change in profit can also represent a change in risk, which directly affects the value of the company. For example, a conservative firm that earned $100,000 annually may have greater value than a firm earning $125,000 if the latter has greater inherent risk in its operation which can reduce the firm's value. Managing financial risk and financial performance is the key to success and survival. Remaining in business relies more on long-range value as opposed to shorter-range profit concerns. Profit planning relies almost entirely on assumptions about the future, while the longer-term approach of maximizing the value of the business relies very strongly on the control of risk and protecting and increasing of the firm's existing value.

Managing a Financially Stressed Company

When a company becomes financially distressed, it does not necessarily mean that it is the end of the organization. Contractors have a number of options to manage and turnaround an underperforming company (Figure 7.2). There are

Figure 7.2 Contractors have a number of options to manage and turnaround an underper-
       forming company.

varying degrees of difficulty, so contractors need to differentiate between a little
slip-up and big financial trouble. If a contractor slips a little, they may take actions
to remedy the situation. But if the contractor is in major financial trouble and
financial losses only seem to be growing, then they need to move the company
into "survival mode" and activate an effective turnaround management plan. The
following sections discuss potential actions to support a contractor in working
their firm through, and hopefully out of, "survival mode."

Banks

It can be tough to get new money out of banks when financial distress is a problem. There is a "big-borrower" syndrome characterized as follows: If you owe a
bank $1 million you have a problem, if you owe a bank $100 million the bank
has a problem. However, in most cases, contractors are not big enough and don't
owe enough to challenge their bank. Bank "workout" programs may be able to
be negotiated, but most are unattractive to contractors because they often depend on showing favorable results in the first year or over some impractical period of time. This is very difficult because turning around an underperforming
construction company is like turning an ocean liner; it takes time from initial
actions to noticeable results, and one year is usually extremely difficult, if not
impossible.

Bonding Companies

It is fair to say sureties provide bond credit to successful construction enterprises and few would take on a new account that was underperforming. If an underperforming contractor has a long-term relationship with their surety, workout is possible but not guaranteed. A positive outcome can only be reached by talking openly and honestly with the surety and as early as possible about the financial distress. However, there is no guarantee what the reaction will be. Obviously, turnaround requires continuing bond credit. Therefore, advising surety early before disaster strikes is the only solution to maintaining a relationship. It is critical to have a viable, detailed plan addressing the required actions to accomplish a turnaround and including any required funding at the initial meeting with the surety. Relying on the surety for funding is not an easy option, and there is little chance the surety will continue bonding. This is particularly the case in light of the reality that the contractor was unable to complete the bonded work at a profit in the past. The contractor must be able to convince the surety of their commitment and capability to complete the unfinished work. The odds of being bonded by another surety, after your surety has incurred a loss, are close to zero.

Outside Support

In survival mode, it is critical to keep control of emotions and come to grips with the difficult decisions required to turnaround a financially distressed organization. This obviously includes partners and top management. Actions necessary for turnaround will appear exactly opposite to what a contractor would consider "normal business practice," which is why it is so difficult for "normal businesspeople" to digest. For this and other reasons, it is probably best to rely on outside advice. Board members or business associates may have experience with turnaround. If not, turnaround consultant advice is probably the only option. While there may be a place for outside law firms in some aspects of the construction business, turnaround is not one of them because time to completion is a primary objective of the surety, while going to court puts the breaks on everything, and even threatening to go to court drives surety to their prima facie right to complete the work.

Bankruptcy

Bankruptcy is a huge step. The upside to filing bankruptcy is that it is an opportunity to cleanse past mistakes. It may work for huge corporations but is often difficult for closely held construction enterprises to manage. A *Yale Law Journal* said that 80% of the companies that attempt reorganization bankruptcy (more in Chapter 11) ultimately fail. A professor at the University of Wisconsin verified that fewer than 20% of smaller companies who enter bankruptcy survive.

Because of the emotional attachment to their company, contractors are often slow to admit and address financial distress, and most lack enough assets to borrow against or sell to fix the problem. It is also difficult to persuade lenders or investors to put up more capital. Bankruptcy dragging on for years can be an incredibly expensive process, and most small companies do not survive the depletion of their value in liquidation. From the authors' experience, bankruptcy is typically not a viable option for small and midsize construction enterprises that hope to restart their company. A contractor who makes it through survival mode will have to take remedial measures and change their traditional business management methods.

Reengineering for the Survival Phase

As discussed in earlier chapters, there are three key functional areas in the construction business. An understanding of the dynamics of the construction business is a prerequisite to taking corrective action during unsatisfactory trends and "survival mode."

1   Marketing and sales – getting the work: In survival mode, this is generally put on the back burner.
2   Operations or production – doing the work: In survival mode, this is usually all that matters.
3   Administration and accounting – measurement of performance: In survival, there is arguably no need to worry about this function; we already know that performance is unsatisfactory. To reverse unsatisfactory performance, timing is critical. Therefore, contractors must concentrate on operations, the only area that can provide cash flow and profit. Profit for construction companies is directly proportional to the attention afforded to field operations.

In the following sections, the authors provide five recommendations for contractors to consider for successful turnaround management.

1   *Serving the Financial Statement*
    When business is going well, a construction firm can be measured by its financial statements. In contrast, during survival mode, financial statements are of limited concern because the firm may not last another year. In normal times, long-term value is more critical than short-term profit. In contrast, during survival mode, to return to stability in fiscal matters, the contractor must operate the business in what may seem like a reckless manner, driving profits first, not long-term value. For instance, a contractor will consume equipment by various means, including using less-than-ideal equipment for jobs to cut costs simply because said equipment is available.
2   *Managing Volume*
    Understanding an organization's optimum volume is critical to success. In rebounding markets, many companies commit to more work and get stretched thin, resulting in less profits. An honest appraisal of an

organization's optimum capacity and capabilities is necessary to determine how much work the company can effectively manage profitably. When existing, trusted, and proven field supervision is fully employed, a construction organization is at capacity. When an influx of work requires promotion of untested people or when new hires are not trained by or are not supervised by trusted employees, an organization has exceeded its capacity, exposing itself to performance risk.

The concept of "volume management" suggests that leadership is responsible for controlling sales within pre-set self-imposed guidelines based on a diligent appraisal of their optimum volume, measured in dollars and/or number of projects. A construction company may grow in three ways: (1) by increasing the number of projects while maintaining project size, (2) by maintaining the number of projects but increasing project size, or (3) a combination of projects increasing in size and number.

With more projects, a company growing by 25% will have a quarter of its production in the hands of untested field management. With larger projects, existing field management may find themselves over their head and perhaps lacking the needed experience. Either way, it is not business as usual, and there is additional performance risk, when compared to work controlled by known and tested employees. If both size and number of projects are increasing, the risk is magnified.

In survival mode, a contractor should immediately cut overhead and downsize, holding onto only their best people and equipment. The contractor will need to reduce equipment payments, slowing them until they are ready for repossession. They may have to institute a moratorium on payables for as long as possible. A 30-day moratorium on payables is, in effect, borrowing one-twelfth of annual payables. This practically acts as a loan to improve working capital.

3   **Building Cash**
Survival activities put bank credit in jeopardy, so cash needs to be protected. In extreme cases, one may consider reducing deposits with banks owed to, and placing new funds with a banking institution where there is no debt. Depending on the severity of the situation, the bank holding the debt may offset the firm's funds on deposit with them. Additional bond credit may not be available but will not be needed for a while because as backlog gets completed it will not be replaced until the downsize target is met. Depending on the depth of the situation, a reduction in volume of 25%–50% may be necessary. If more than a 50% reduction is indicated, the firm may not be worth saving.

4   **Reducing Staff**
Reducing staff is arguably the most difficult decision, especially for small to medium-sized companies. At the same time, if staff is not reduced, the company may bankrupt and all employees will lose their jobs. One recommendation is to not lay off the field staff early on, in part because the projects need to be completed. New work is not an issue as downsizing is a primary element

of survival. Overhead costs need to be reduced and staff will be the primary area of layoffs, unfortunately. With the reduction in size, fewer marketing personnel, if any, are needed. The same goes for accounting and administrative personnel. Morale will be an issue so it is critical that leadership in the office and in the field explains to the remaining employees how the recovery plan will work.

5   **Reducing Assets**
    Cash needs are primary, and while there may be the inclination to sell off the old equipment, that does not generate much. Selling newer equipment raises more cash. Replacing equipment with lease or rental also improves cash flow. These are drastic but often necessary moves in survival mode because selling assets generates much needed cash, reduces debt, or both.

Liquidity Management

Conservative liquidity management provides fundamental protection from financial distress and should be included in the business philosophy of every construction company. It should be a part of a contractor's DNA (Figure 7.3). The idea of having some cash or credit reserved for a time when it may be critical but unexpectedly needed is not a standard element in construction business accounting. However, as earlier mentioned, for an industry with the second highest company failure rate in the country, it should be. Many of the hundreds of failed

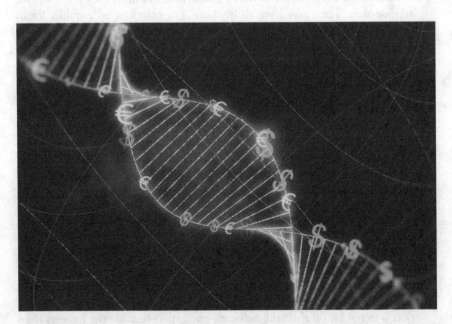

Figure 7.3 Conservative liquidity should be a part of a contractor's DNA.

construction firms the authors have analyzed would not be out of business if they had set aside a rescue fund to save their company from financial disaster.

The authors advocate for liquidity reserves as a necessary element of risk control in a prudently managed construction business. The recognition, measurement, and management of risk are critical skill sets in the construction industry. However, even with consistent application of these skills, mistakes still happen and there will be occurrences outside of a contractor's control. When all else fails, cash reserves are the final line of defense. Simply put, you would probably not board a ship if it did not have lifeboats; then, why would you operate a high-risk business without some reserves?

Some will argue that warehousing liquidity will slow the growth of a company and would be better reinvested in the business. However, total reinvestment means being continually one misstep away from financial distress. It only takes one horrific job, mistake, unforeseen crisis, or market turnaround to push the overextended company close to the edge. If you have been in business for a while, you probably already know of someone who has faced this scenario. It can happen to any organization, and conservative liquidity management offers protection.

Financial Risk Management

Liquidity reserves offer a chance to reduce the risks in your business. The primary considerations in the measurement of risk are the likelihood of the event occurring and the magnitude of the penalty if the event occurs. The likelihood is commonly measured in odds, such as "the odds of being struck by lightning are a million to one." However, while the odds are extremely in your favor, the penalty is too high. Therefore, no one knowingly puts themselves in a position to be hit by lightning. Consider this other example: the odds on the flip of a coin are 50/50. In this case, if one was to bet a dollar, both the odds and penalty are reasonable. However, if one was to bet a million dollars, the odds remain reasonable, but it is now an extremely high risk that few would take.

You cannot measure risk without evaluating both the odds and the magnitude of the penalty or reward. For example, if the odds were six to one in your favor, would you take the bet? This question cannot be answered without knowing what the event is. If it is a ten dollar bet, then it is a bet that most people would take, because of the excellent odds and nominal penalty or reward. However, the same odds exist in a game of Russian roulette. With one bullet in a six-cylinder revolver, the game requires the player to pull the trigger aimed at his or her head. Good odds, but much too severe of a penalty.

For the skilled construction professional, the odds of mistakes such as occurrences outside your control or just bad luck are fairly small. However, some of these things happening with bad timing could put a business into financial distress, which is obviously a serious penalty. In construction, we face these risks every day and do not get to opt out of them. However, we do get to hone our skills at recognizing, measuring, and managing risk. One critical risk control option is establishing liquidity reserves to counter the eventuality when all else fails. Using this approach, a contractor can build a safety net, a protective shield against financial distress.

Professional high-wire performers never intend to fall, but they still want a safety net below them. Most contractors work without a financial safety net. Not that long ago, historically high margins (rewards) may have made that a good bet. However, today's slim margins have shifted the risk/reward balance to the extent that protection against failure is practically a necessity.

Measuring Liquidity

The amount of liquidity reserve will vary depending on the contractor's appetite for risk and the amount of liquidity they are comfortable with (providing the ability to meet short-term obligations). While it differs for each company, an average comfortable liquidity is when the debt to worth ratio is around 1 to 1 or less, meaning that the company can settle its liabilities using its assets; while its net quick is around 1.5 to 1 or more, meaning the company has the ability to settle short-term obligations with assets that are easily convertible into cash. It is best to use an average from the prior three year-end financial statements as a benchmark, assuming the company was profitable and increasing in value during those years. If not, we recommend going further back until there are three consecutive years of increasing value. When this occurs, there will have been cash, cash equivalents, or unused credit available. The average of that amount, over the three years, is what can be referred to as "liquidity reserves." Liquidity reserves can be defined as the funds that are available to the contractor to pay the bills in the event of unexpected financial difficulty. Most construction companies are liquid in good times, but only prudent contractors remain liquid in bad times. These prudent contractors will pay their bills on time during financial stress because they have a liquidity safety net against failure.

If the potential of a financial decision exposes your life's work and net worth to disaster, is that a risk you want to take? We are not implying that the successful construction companies need to be perpetually risk-averse; however, we see no advantages in betting everything. It is like playing Russian Roulette with your livelihood. We have seen too many contractors and their families lose it all, and we are certain that prudent management of liquidity is a winning long-term strategy for the contractor of the future.

Final Thoughts

During the survival phase, the contractor must concentrate on first-year goals and show positive results as soon as possible. Showing the surety and bank even modest performance improvements is important because it demonstrates the recovery plan is heading in the right direction. If the turnaround process is verified, there is a possibility of reduced pressure from stakeholders.

No construction company goes through survival twice. They do it once, and, if it works, they typically never get caught in underperformance again.

This chapter explored challenges faced during growth markets, key indicators, and metrics for assessing financial distress and managerial techniques supporting financially stressed construction companies. In the next part of this book, the authors will share key business management information that is critical to a construction company's success. The next chapter will further expand on risk identification and risk management.

Review Questions. Check All That Apply.

1 A proven defense strategy during a declining market includes:
   a   Cutting the cost of doing business
   b   Downsizing
   c   Cutting overhead
   d   All of the above

2 In "survival" mode, the most important functional area is:
   a   Marketing and sales
   b   Operations and production
   c   Administration and accounting
   d   All of the above

3 A true measure of a construction firm's performance is:
   a   Increase in gross profit
   b   Increase in sales
   c   Increase in value
   d   Increase in assets

4 Turnaround management includes:
   a   Building cash
   b   Reducing assets
   c   Serving the financial statement
   d   All of the above

5 Liquidity reserves over a chance to:
   a   Increase assets
   b   Reduce assets
   c   Reduce risk
   d   All of the above

Critical Thinking and Discussion Questions

1 Explain the "compression" dynamic.
2 Explain the concept of "value versus earnings."
3 Discuss the author's recommendation for managing a financially distress construction company.
4 Describe the author's recommendations on financial risk management.
5 Explain what the authors mean by a "comfortable liquidity."

Part III
Managing the Business

8 Recognizing and Managing Risk

Characterized by tight margins, periodic labor shortages, and competitive new technologies and methods, the construction industry operates in a landscape that has little room for error. The construction market is continuously cyclical with downturns that can weaken some construction organizations to the extent that they have difficulties financing their growth during market recovery. These challenges increase the probability of business failures and contract defaults.

Now more than ever, owners, contractors, and designers need to improve their risk awareness and risk protection capabilities. They should be concerned about the fact that construction business failures are worse in market recoveries compared to market slowdowns. At the beginning of a downturn when work decreases, the balance sheet blossoms as old receivables continue to come in and less money goes out for job costs. The opposite occurs during growth periods because the cost of producing more and more work each month exceeds the receivables from the prior months. Cyclical market downturns weaken some construction organizations, to the extent that they have difficulty financing their increased workload and growth during recovery, which increases the potential for defaults. When this occurs, it is possible for contractors to "cash flow" themselves out of business.

Risk exists in any commercial transaction, but the construction industry is particularly vulnerable to risk. Construction risk is sometimes poorly defined and often misunderstood; therefore, the responsibility for various project risks can be ambiguous. Designers avoid risk, owners prefer to pass it along, and contractors absorb it. To contractors, "risk" is not a bad word. Thirty years of research into the causes of contractor failure uncovered a noteworthy truth: that assuming risk is part of a contractor's DNA.

Risk Factors

Today's construction risk environment is dramatically different than it was in the past, and risk factors are mutating just enough to be almost unrecognizable. In addition to project risks, inherent construction risks can result from changes in the contractor's typical project size, type, geographic area, key personnel, and/or managerial maturity. The cyclical nature of the construction market also magnifies

DOI: 10.1201/9781003229599-11

the risk of project and contractor failure. These risks are most pronounced and arise regularly in a recovering market:

1   **Shortage of Skilled Labor Risk:** Following a construction market decline, a number of workers leave the industry. Therefore, when the industry starts to grow again, companies are short on skilled labor and may not be able to complete quality work on time and on budget.
2   **Company Growth Risk:** After an economic recession, the market begins to grow again. At such a time, growth is always welcome and rarely seen as a major cause of impending failure. However, research reveals when a construction company expands in size, it requires careful management decisions to reduce the risks inherent in changes. Growth itself impacts failure and the potential for financial distress.
3   **Subcontractor Risk:** Subcontractors are a fundamental part of the construction process and are subject to most of the same market changes as general contractors (e.g., labor shortages, material price escalation). Each subcontractor is critical to a project's success, so it only takes one disruption in the entire process to intensify project risk (Figure 8.1).
4   **Capital Risk:** Secured equipment loans, unsecured working capital credit, and surety credit for bid, payment, and performance bonds are vital to the growing construction enterprise. Long durations of market downturns can financially weaken some companies and impact their credit worthiness. A construction company may show positive income on its financial statement, and then suddenly experience a financial crisis due to a lack of cash and limited borrowing power. A considerable number of construction companies exit the business because they run out of credit.
5   **Commodity Risk:** Construction is a custom service and product that some project owners are beginning to think of as a commodity, causing some to believe that all construction companies perform alike. This, in turn, encourages

Figure 8.1 Each subcontractor is critical to a project's success.

owners to be less discriminate about contractor selection and to think of price as the single key differentiator. This is one of the reasons profit margins are low, when compared to the historic norms. When buyers of construction services believe a product is a commodity, they generally expect to pay less.

6   **Contract Risk**: In this highly competitive growth market, contract terms concerning the responsibility of each party to the agreement are changing. Construction buyers (owners) and designers attempt to shift risk in varying directions. At the same time, case law (court rulings concerning construction disputes) sometimes further cloud the issue of which party is responsible.

7   **Change Risk**: When a firm expands in size, it gradually evolves into a different organization. Change always has risks associated with it. For example, growth-related changes impact the amount of capital required, the time and attention management can spend on multiple projects, and the expertise required to complete new types of projects.

8   **Accounts Receivable Risk**: Accounts receivable are an asset but unfortunately cannot be used to pay bills. There are any number of failed construction enterprises that may have survived if they had all the money that was owed to them. Poor management of accounts receivable including managing the payment process is one of the major risks in the construction business.

9   **Innovation Risk**: Advancements in field and office technology, and developments in project funding methods, introduce potential risk initially. Selecting and implementing new technology can significantly consume management's time and attention; if the selected innovation does not perform as expected, that time investment can be costly. New methods (e.g., construction methods, public-private financing) are initially risky simply because project stakeholders have limited experience with them. The authors are proponents of innovation and work in this realm, but recommend that any related risks be identified upfront to make informed decisions.

Considerable attention from management is necessary as the industry continues to evolve with an increase in new and sometimes unrecognizable risks. In the next sections of this chapter, the authors will discuss methods of risk management and risk mitigation.

Early Signs

While most owners, construction managers, designers, contractors may complain when a project goes bad, many express some variation of *"I had a bad feeling about this for months"* or *"All the signs were there, but I was hoping for the best,"* so they may not always be totally surprised. Some early signs of trouble before default include (1) complaints from suppliers or subcontractors about unpaid invoices, partial or late payments, (2) contractors asking for advance payment or help making payroll, (3) unexplained cuts in crew size, (4) declining work quality, (5) overbilling of quantities or percentage of work completed, (6) unusual requests for payment of materials ordered and stored off-site, and (7) changes in foremen, supervisors, or middle management. Long-term employees can often sense trouble early on and may leave for other opportunities.

Protecting your organization against default is much easier said than done, since several commonly recommended protections could also cause problems. For example, one recommendation can be summarized in the saying, *"Don't take a low-ball price."* However, if a number is on the street, someone else is going to use it, and you might not get the job if your bid is higher. This said, one needs to fully consider the additional risk involved and figure out whether it is really worth it. In good times, this is obviously less of an issue, since aggressive pricing is not as prevalent, and, if you don't get the job in question, there are other opportunities available.

Managing Risk

Management decisions generally determine whether an organization will succeed or fail in the ever-changing construction business environment (Figure 8.2). The decision-making process begins with management "beliefs" that must be regularly reexamined as the business environment evolves. Beliefs that were appropriate in the past may not be relevant in the present or in the future. Some unexamined beliefs that have been accepted in the industry for a long time are no longer valid, such as *growth is always good, having some unprofitable work is unavoidable,* and *past success implies future success.* These beliefs should be reevaluated.

Figure 8.2 Management decisions generally determine whether an organization will succeed or fail.

Construction is a highly complex endeavor that is worked out over a relatively long period of time. Its success or failure is affected by variables including weather conditions, labor problems, inflation, unexpected rises in interest rates, the high cost of equipment, a tightening or shrinking of the market, or simply bad luck. This variety makes measuring a project's risk in advance difficult and requires considerable knowledge about the construction enterprise and the current construction environment – a specialized field in itself. A contractor may seek information and experts' perspective in risk management by engaging with their internal and external accountants, attorneys, insurance, and surety partners.

Risk in construction cannot be fully eliminated, but it can be mitigated. However, risks cannot be mitigated until they are identified, measured, and thoroughly understood. This is complicated because the various parties involved in the construction process see their respective roles in addressing risk differently. Many construction risks are attributed to more than one entity, making identification difficult, elimination almost impossible, and mitigation the only viable alternative. The three key risk management steps for owners, contractors, sureties, bankers, and designers consist of:

1   Recognizing and identifying specific risks in advance
2   Assessing and quantifying their importance and exposure
3   Mitigating and managing their impact and cost

To accomplish these steps, specialized knowledge, industry intelligence, and experience with large numbers of similar projects are required. It is also necessary to have access to some hard-to-find information, some of which is very confidential. Sureties have extensive experience and data from pre-qualifying firms in all types of construction and from various locations, but the data are not easily accessible.

The willingness to take risk is at the core of the construction enterprise. Every time a contractor signs a new contract, they voluntarily assume risks that are not always fully defined. The ability to recognize the true nature of risk, assess its impact on an organization, and take steps to mitigate those impacts will be a fundamental skill set of successful industry stakeholders. A partnering of owners, contractors, sureties, bankers, and designers can play an important role in successful risk mitigation.

- **Owners** have one of the easiest means for risk mitigation, as they can bond around their exposure and have a knowledgeable surety to complete the critical pre-qualification process. Sureties can screen out ill-equipped enterprises, and they cover the cost if it does not work out.
- **Contractors** that are successful over the long term typically elevate risk management within their organization and embrace formal risk management processes, which fortunately are becoming the construction industry's newest discipline.
- **Sureties** have an important role to play in the construction process and bring a unique perspective to the contractor and subcontractor screening process.

- **Bankers** can have the same protection as owners by requiring payment and performance bonds on the projects they finance.
- **Designers** can assist owners with the selection of contracting methods and project participants, while encouraging collaboration among all parties of the construction process.

The successful contractor of the future will need to learn an entirely new skill set to recognize risks hidden in the market, management decisions, and economic climate. What looks like good news, too often, has the potential to be hazardous. Diligence is required because construction business and project risks are often disguised in a variety of forms. For example, top-line growth may appear to be an avenue to success but can also lead to failure. Tightly drawn contracts and pushing risk toward others may appear to be protection but can create more problems than it solves. An expanding market may look like an opportunity but can also be a minefield. Successful contractors are those who react quickly to evolving market conditions.

In the construction business, past success is not an indicator of future success. In fact, the authors' research on the causes of construction business failures indicates every major change in a successful organization, particularly in growth periods, creates a period of risk despite previous successes. When a construction organization substantially increases in size, it is no longer the same company it was before, and may not be successful if it maintains pre-growth management methods.

The successful construction enterprise is organized to be market-driven, and not volume-driven. It strives for carefully planned growth but is prepared to level-off or fall back on volume if the marketplace shrinks. The company uses its markup as a competitive tool and does not take break-even work to maintain sales. Flexible overhead (explained earlier in this book), if built into the organization, can be cut immediately when not needed, and permanent overhead should be cut in downsizing, as needed, so that the contractor can fluctuate in size to adjust to changing market conditions.

Evolving Markets

Companies succeed and others fail, both in good and bad markets, in part as a result of how companies adapt to continuous industry changes. Consider the following example: a general contractor was underperforming because of dramatic changes in the marketplace they served. The contractor was advised that the construction industry was consolidating and that they needed to either get bigger (possibly by acquiring another company) or smaller (possibly by focusing on their core business only, or by being acquired by a larger company) to reverse their decline but, unfortunately, could not be convinced. Over several years, a number of their competitors grew substantially or merged with larger firms, and during the same period a few of their best customers were bought by national firms and began hiring larger contractors for construction and maintenance work. The market they dealt in was changing, and the availability of their typical-size work was shrinking. With the contractor's limited resources, getting much bigger was unlikely. The founder of the company insisted he did not put his entire life into building his company to get smaller now. His refusal to react eventually cost him his business because they could not compete effectively in the new market. Many other contractors are facing similar circumstances leading to some dramatic changes in the construction industry. Some predict that many of today's midsized companies will be merged, acquired, or otherwise cease to exist over the next 10–20 years.

Contractors are receptive to information about performance improvements and profitability enhancement but are often resistant to research defining consolidation within the industry. Most contractors will not react until there is a definite certainty. Construction is rapidly becoming standardized, technology simplifies the process, and growth in typical project size are a few of the significant changes happening all at once.

Not that long ago, contractors built custom wooden forms from scratch for complex concrete projects. The competition was limited to contractors with the know-how to complete this specialized work. But now the process has been revolutionized by patented form systems which require way less specialization to install, making this work accessible to a lot more contractors. Standardization modifies an industry, encourages new entrants, and helps lower costs due to economies of scale.

Historically, construction was a "can-do" industry of brawn and might, with a tradition of building projects with basic equipment and limited technical support. A lot of projects appear to be built with a backhoe and a box full of tools. However, the truth is that, in the past, success demanded a rare combination of talents, including the ability to muster resources to build a project; putting an accurate price on the work in advance; managing labor, subcontractors, vendors, and designers through a long and arduous process; and tolerating a high degree of risk. Contractors of this breed had great success, as long as their methods of bidding, especially their costs for individual items of work, were closely guarded secrets. But today that has changed. New technologies and wide access to information have demystified the processes of estimating, organizing, and producing the work. Even buyers of construction services and designers have become quite knowledgeable about such processes and the underlying costs.

Construction was a custom effort. Double-digit profit margins compensated for the hard work and some inefficiencies. Now contractors are working just as hard and making less profit. Even in an exceptionally good economy, construction margins have typically hovered in the mid and low single digits, and even lower for larger contractors. Today's structurally altered margins will not return to historic highs. Consequently, as construction becomes increasingly standardized, efficiency and productivity will become the major differentiators among competitors.

Some pass up the chance to become more efficient, with excuses including that they can't afford to upgrade and are too busy for training. Low margins restrict resources, making it harder for small and midsized businesses to invest in increasing efficiencies. Larger companies suffer the same low margins, but their size enables them to invest in production improvements to satisfy the shifting demands of owners. Continuous improvements have become a necessary cost of doing business and are becoming critical to survival.

Ironically some small contractors may fare better than some midsized firms, because there will always be a place for niche players. Specialty contractors enjoy relatively higher margins until consolidation moves into full throttle. Small and midsized firms may be able to accomplish large-scale efficiencies through cooperation, but their window of opportunity will be brief. Profitable and well-managed enterprises will survive by becoming part of larger organizations or by serving niche markets.

Payment and Performance Bonds

Another important practice for risk management and mitigation relates to effective surety bonds. When working with unfamiliar subcontractors, bonding them is often good advice, even if not required, especially if their price is lower than any of the others. However, when a prime contractor tries to pass along the cost of a bond not required by the specifications, they risk outpricing themselves and not getting the job.

Prime contractors may find low-ball offers to not be bondable, and some move forward anyway, particularly if "promised" a bond. In some cases, project teams learn after construction is in progress that the subcontractor could not secure a bond, hearing excuses like "*We tried to bond the contractor, but they could not get a bond*" or "*We were promised a bond, so we issued the contract.*" The ability to secure a bond with a subcontractor may be an indicator of a contractor's diminishing reliability. Holding back payments is a common solution; however, it often compounds cash flow issues and accelerates the potential for default. Joint checks have the same impact.

In the past, the cost of a subcontractor's default was the payment of unpaid vendors and subcontractors; however, in today's construction industry, there is an increasing loss exposure from defective work. Organizations with serious quality-control processes and those that foster a quality culture are able to prevent or mitigate these losses.

A contractors' financial condition will obviously deteriorate if there is poor management; however, in today's reality, the failure of one subcontractor on a project can have a significant impact on the financial condition of other subcontractors on the job. Smaller subcontractors may only be as secure as their last job. Because pre-qualification isn't an absolute guarantee against default, contractors may need to learn a whole new skill set, including how to manage defaults to minimize exposure.

Late Payments May Be Your Own Fault

One way to reduce the risk of default is managing cash flow and ensuring contractors are paid on time. Late payment to contractors is an industry nightmare that violates contracts and multiplies risk, but, worst of all, some contractors allow it to happen (Figure 8.3). Having been paid late for so long, some contractors think that it is "normal" and most feel they have no influence over the payment process. We call it an "entitlement" paradox because in a commercial transaction, payments are earned, not bargained for. If contractors assent to not being paid in accordance with their contracts, they share the blame by not demanding what they are entitled to. Not getting paid on time takes some of the fun out of the construction business and adds to the risk.

Despite contract requirements, late payment in the construction industry is commonplace. The contractual arrangements guaranteeing and assuring the performance of the contractor and payment by the owner can be one-sided, giving the owner more protection than the contractor. On public work and, to a lesser extent, on private work, 100% payment and performance bonds are required. This effectively assures the owner that if the contractor is unable to finish the job at the agreed-upon price, the bonding company will complete the project and pay unpaid labor and material already in place. However, contractors have little assurance that the owner will have the funds and the inclination to pay in a timely manner. Some organizations have entered into contracts and then were unable to pay for them, and many contractors on public projects have performed signed change-order work and could not get paid because the owner or their representative did not have the proper authorization.

Figure 8.3 Late payment to contractors is an industry nightmare that violates contracts.

Retainage: Impact and Cost

Even with 100% surety-guaranteed protection, the project owners usually also withhold part of every payment as further guarantee that the contractor will perform the job. Retainages of 10% are common, with some retainages reduced to 5% when a job is 50% completed (and acceptable). The federal government has begun to deal with the inequities of huge retainages on their projects, with some progress to date. Retainages are an expense to a contractor, which are hopefully passed on to the owner by inclusion in the price of the work.

When retainage is included in contract language, the assumption seems to be that the contractor will not perform his or her job unless they are forced to. Project owners and contract authors often deny any built-in bias or mistrust, but they are reluctant to give up the practice. Its usage implies distrust in the relationships between owners, contractors, and designers. This can impact the environment in which contractors work. Trying to get retainage released at the end of a project from designers and owners further perpetuates a strain on the working environment. When one million dollars' worth of retainage are held for several months awaiting 50,000 dollars' worth of punch list work, it becomes obvious that mistrust exists and that existing contractual arrangements provide financial advantages to the owner. Owners are earning interest on the million dollars retainage, while contractors are paying interest on loans to compensate for high retainage.

Payment Requisition: Contractor Entitlement Issues and the Designer's Role

This environment causes many contractors to feel they have little power to collect their payments and that entitlement to their money is somehow clouded. Billing procedures in the construction industry are unique. Contractors earn their money, but too often they don't seem to have a strong feeling of entitlement. This condition is reinforced by the practice of walking a site with a designer or his/her representative at the end of a month to agree on the percentages of completion of the various line items of work. The designer's representative and the contractor's representative usually have a copy of the last month's payment requisition (application for payment) in front of them. They either agree or debate what was completed that month. This practice is common on unit price jobs, but also prevalent on lump sum projects. Some contractors send in their completed requisition, and if the designer disagrees, the designer sends it back marked with changes, and the contractor retypes it, signs it, and sends it back to the designer. These practices display the payment environment in the industry and illustrate a problem – the process puts contractors in the passive role of taking what they can get rather than the active role of invoicing their customers for services rendered.

The very word "requisition" suggests that contractors can only "ask" for their money and that there are questions regarding their entitlement to it. Contractors claim that this cooperative approach is the only way to get the requisition approved by the designer and into the hands of the owner for payment. Some contractors rely on the architect or engineer to treat them fairly. Under the facade of fairness, contractors have tolerated the attitude that they need to be controlled and that the designers must protect the owners from contractors. Fairness is a two-way street. It is not something that a party to a contract must ask for – they are entitled to it. If a contractor does not get paid on time for what they have earned, it might be their own fault for not demanding what they are entitled to according to the terms of their contract.

Most construction contracts provide for the payment requisition to be filled out by the contractor and approved (or disapproved) by the designer in a specific length of time. A contractor does not need outside help determining what work was performed during a month nor how much should be paid for it. If the designer does not want to approve it, that is their prerogative. They can red-pencil it and send it on to the owner along with the amounts they think should be paid. There is no contractual requirement for a contractor to retype their original requisition any more than a need to walk the site with the designer and bargain for the amounts. All that is established by doing those things is to set a precedent, implicitly agree to it, and potentially modify the contract (by adherence) that the designer decides what the contractor will be paid. Retyping means changing the requisition to the amount the designer approved. By doing this, the contractor places their monthly payment amount in the hands of the designer.

Effectively what the contractor has done is establish that they and the designer will agree on the amount before it is approved and sent to the owner. This not only rewrites the payment provision of their contract, it also puts the contractor in the one-sided position of trying to talk the designer into agreeing on how much should be paid. However, the designers do not have to pay the construction sub-contractors and suppliers, nor have they already paid for labor for the month. The designer is a third-party as far as these aforementioned costs are concerned, and does not have the same time constraints. So why do some contractors get involved in these practices? Some say, "*It's easier than arguing over payment, because arguing will only cause other problems with the designer.*" Others have said, "*I don't want the owner to see a red-penciled requisition, because they will think I was asking for more than I should have.*" And still others say, "*The owner is not going to pay except for the amount the designer approves anyway, so what's the difference?*"

The practice gives the designers a disproportionate amount of control over the contractor during the construction process. If a contractor intends to live up to their end of the contract (including every item in the plans and specifications which they have agreed to perform anyway), the contractor has nothing to fear from the designer. There is no need to shelter the owner from payment requests. According to most contracts, the owner has an active role in the construction process, and one of the owner's responsibilities according to the contract is to pay for the work. It is appropriate to clarify the necessity of the active involvement of the owner in the payment process. The authors strongly recommend cooperation on sensible and businesslike terms but not one-sided cooperation where the contractor assists the designer in sheltering the owner from effective participation in the two-party contract between the owner and contractor. Contractors are not powerless in the payment process.

Case Study

Next is a case study from a contractor's experience. Please note the authors are not recommending that the readers take similar actions; the case is for illustration purposes only.

> The first requisition on a project was sent back to me (the contractor) from the architect, red-penciled with reduced amounts and a note to retype it and send a clean "original" back to the him (the architect). I proceeded to send the owner my original requisition, the red-penciled copy, and the architect's note to me. I included a detailed explanation quoting the terms of the contract that said the owner had to pay the amount approved by the architect. I went on to explain that I disagreed with the reductions and could demonstrate that each line item was correct as of the last day of the month requisitioned for and that I would be invoking the dispute resolution clause in the contract to be paid the reduced amounts plus interest. I copied the architect with the letter and attachments.
>
> Shortly thereafter, I received a call from the incredulous architect screaming that their firm never sent an owner anything but a clean typed requisition. When an explanation was demanded I responded that as this was the first requisition on the project I could only assume that he (the architect) had reduced most of the quantities, out of habit or perhaps because he thought he should, but that the amounts were all correct and, if anything, understated; and that one line item requested zero where anyone could see that the work was in place by the last day of the month. The architect said that he wouldn't send the red-penciled copy to the owner, and I would just have to change it, retype it and sign a new one. I told him that I never retype a requisition, that what he had was a copy, and that I had sent the owner the original requisition, his red-penciled copy, and a copy of his letter telling me to change it--and that I was waiting to see what the owner wanted to do.
>
> The architect then proceeded to inform me that the contract stated that I could have no contact with the owner, except through the architect. I told him that he would have to explain that to the owner. He said, What if I send the owner the copy of the marked-up requisition and tell him not to pay any of it? Both he and I knew he would have a hard time explaining that, and it sounded too much like discipline for me to even answer. He hung up, called the owner, and advised them not to pay anything. When the time ran out and the payment became overdue according to the contract, I gave three-day notice that the work would stop because the according to the contract the owner was in default for non-payment. That of course brought everything to a head and an immediate meeting was called. The architect quoted the contract explaining he was in charge of everything, but the owner's attorney clarified that the architect was not party to the contract.
>
> After much discussion, the architect was asked if he could document that the work in question was not completed by the last day of the month; at

which point the architect said that he visited the site on the 6th of the following month and speculated from observations what may have been completed by the last day of the prior month. Then, I was asked how I arrived at the requisitioned amounts and replied: I was on site the last day of the month and took measurements and pictures. When pressed by the owner for details the architect said: I'm not sure exactly what was done by the last day of last month, so I used calculations. This ended the discussion. I was paid immediately and before the tenth of every month for the rest of the project. Was the inspection on the job any more difficult than usual? I'm not sure. But I was sure that I wanted to be paid the correct amounts I had earned, and on time.

Invoicing Issues and Timing

The authors recommend that contractors stand firm on contractual payment requirements from the first requisition and according to the contract. If you do not get treated in a professional manner when it comes to getting paid, it may be because you have not acted professionally, and therefore cannot demand businesslike treatment. Contractors are expected and required to perform professionally, live up to their contract obligations, and build in accordance with the plans and specifications. As long as they do that, they have a privilege to demand their right according to the contract, including getting paid on time.

Slow pay and retainage held too long are an increasing problem in the construction industry (Figure 8.4). Too many people are allowing it to get worse by saying, "*We can't do anything about it.*" Each contractor must decide for themselves what their approach to this subject should be after considering the impact of slow pay on their business. Interest paid on lines of credit required because of overheld retainage adds up quickly. If contractors were paid for the work on time, more of them would be out of debt.

One of the things a contractor can do to collect his or her money faster, if not on time, is to prepare and send out their requisitions on time. Some requisitions go out on the 5th, the 6th, or even the 10th of the month when the contract often states: "*Bill for work completed as of the last day of the month, to be paid no later than the 15th, (20th, 25th or whatever) of the following month.*" Designers and owners cannot be expected to begin the payment process until they receive the payment application.

A contractor should begin the billing preparation well before the last day of each month. They should have a tentative requisition completed by the 26th or 27th, anticipating work that will be completed by the last day, and verify it with the job by phone. The requisition should be sent using the fastest method on the 1st of the month, even hand-delivered if practical. They should call to make sure the bill has arrived and is not sitting on someone's desk or in their email inbox. It is best to let all concerned parties know that prompt and proper payment is important to you. Most respect that because it is a sign of a good businessperson. They should not start pushing for payment on existing projects. Instead, the process needs to start from the very first requisition on the next job, and it should

Figure 8.4 Slow pay and retainage held too long are an increasing problem in the construction industry.

be ensured it is consistent after that. It is hard to change attitudes mid-job after everyone has gotten used to a way of doing things.

There is nothing wrong with expecting to be paid for the work and asking to be paid on time. The process must start at the initial project meeting. After all parties express their expectations or needs, the contractor should voice his or her needs as well. The contractor should upfront and say, "*I want to talk about the payment process.*" It's not going to be important to anyone unless the contractor makes it important by letting everyone know they are not embarrassed about it and expect payments to be processed on schedule and in accordance with the contract. Payment should not be a "backroom" or confidential discussion. The contractor should explain that their efficiency and productivity depend on paying the subcontractors and suppliers on time and that they want to build the project, not "invest" in it. The designers and owners know that. They should not need to be reminded, but some do. If you are a contractor, manage the process and keep it in the forefront of everyone's mind. No one else will do it for you.

A contractor should clarify the payment procedures before the first requisition is due and make their needs and expectations known. The first requisition should be pristine, exact quantities of work completed, accurate to a fault, even under a little is fine. This way, if there is a problem with the designer, the contractor can proceed with confidence right from the start, knowing they will be proven correct if the pay request is red-lined. Designers will become accustomed to trusting the contractor's word for work completed and will earn a reputation for fair billing. It is a proven strategy.

Contract payment clauses often state that payment is to be made by the 15th or 20th day of the month, or so, but many owners take that to mean that it is not "due" until the 15th or 20th. A consequence of thinking payment is not due until a certain date is to assume that it is not overdue until a grace period has elapsed, perhaps 10 or 20 days after the due date. Some owners may feel that way, but "by the 20th" means any time before and up to the 20th, not after the 20th. It will only continue to mean that if contractors insist on it in order to get paid on time.

Consider this short case:

> I (*the contractor*) was talking to a public official about getting payment for a current monthly requisition on a project. The official told me the requisition was approved by the public body but that he did not have the authority to pay before it was due. We got the contract out and reviewed the payment section which said, by the 20th of each month. The official asked me 'Then what is the due date?' I replied, according to the contract, the payment is due upon approval of the designer, but no later than the 20th of the month. Without another word he authorized the check.

The authors are not suggesting that getting paid is easy by any means, but instead they are proposing the need to be confident and assertive in claiming one's right according to the agreed-upon contract. Furthermore, we should read and understand our contractual payment provisions and explain them to the owners and designers if necessary.

Cost of Late Payments and Impact on Cash Flow

The cost of accepting late payments is obvious, but the cost of late retainages may exceed even the interest paid for the delay. If a contractor has cash flow issues and the amount of retainage far outweighs the value of the missing items, a lot of contractors give away value in order to get paid. They may often perform work they are not responsible for because arguing about it takes time and delays payment collection. Many contractors have been forced to negotiate forgiving part of the retainage as refunds for incomplete work or quality complaints. Oftentimes that strategy is unfortunately used as leverage to expedite payment of retainage, forgiving some amount in return for collecting the rest. The forgiven amounts are supposedly for work not performed, but are often pure concessions. This can happen even on jobs where a contractor cooperates throughout the project.

Cash flow is always a concern, but when a contractor's marketplace weakens and his or her work slows down, cash flow can become critical very quickly. If all of a contractor's reserves are tied up in their receivables, they can be forced out of business. If many of the receivables are from outdated retainages and overdue payments, the contractor must shoulder part of the blame for not demanding what was rightfully theirs. If they have fulfilled all their responsibilities and the owner does not fulfill their end of the contract, a completed facility is excellent collateral. The contractor should not go under because someone is failing to honor a contract that they have completed.

Conclusion

Construction is a complicated business that should only be attempted by owners, contractors, and subcontractors with strong capabilities in risk recognition, management, and mitigation. The nature of construction risks is changing every day, increasing the potential for business and project failure. Historically, risk management was considered by many to be an insurance issue; however, there are significant business risks not addressed by insurance. Some hidden risks such as skilled labor shortage, limited capital access, payment issues, and adaptation to a rapidly changing business environment have rendered traditional insurance tools and risk management approaches insufficient.

The "successful contractor of the future" will establish formal risk assessment processes and protocols and adopt a strategy of flexible overhead that can easily adjust to a cyclical construction market that invariably presents new risks. The successful owner of the future may choose to transfer risk by relying more heavily on sureties to pre-qualify contractors and provide financial protection against defaults, and will partner with the contractor to adequately address project risks. Prime contractors and construction managers can get similar surety protection for subcontractor defaults.

When the construction market changes from contraction to expansion, which happens about every decade or so, unexpected and unrecognized risks are introduced. In this environment, hanging on to old beliefs will be a dangerous mindset. There are few choices when it comes to risk: assume it, manage it, or transfer it. Construction companies that have not demonstrated proficiency at recognizing, managing, and mitigating risk have a higher likelihood for business and project failures.

Market changes can also be experienced when a contractor, in a new target market, decides to expand or chase leads outside of their regular geographic business areas. Expanding into a new market may be an effective opportunity to grow a business; however, there are numerous unforeseen risks throughout this process that contractors must be capable of dealing with. New market entry is further discussed in the next chapter.

Review Questions. Check All That Apply.

1   The author's list of early signs of trouble before default include:
   a   Asking for advance payments
   b   Increased competition
   c   Equipment depreciation
   d   None of the above

2   Key risk management steps include:
   a   Recognizing and identifying in advance
   b   Assessing and quantity importance and exposure
   c   Mitigating and managing their impact
   d   All of the above

3   Who do the authors recommend should be the party to improve the late payment issues?
   a   Owner
   b   Designer (architect or engineer)
   c   Contractor
   d   Bank

4   Who does the surety bond protect?
   a   Owner
   b   Designer
   c   Contractor
   d   All of the above

5   Who do the authors describe as having a disproportionate amount of control over the contractor during the construction process?
   a   Owner
   b   Designer (architect or engineer)
   c   Lender
   d   Bonding company

Critical Thinking and Discussion Questions

1   Discuss the early warning signs before default.
2   Contrast the construction industry described in the past and present.
3   Explain the roles of the contractor, designer, and owner in the monthly payment process.
4   Explain what the authors refer to as the payment requisition "contractor entitlement issue."
5   Discuss the cost of late payment and its impact on cash flow.

9 Entering New Markets

Entering a new construction market is a complex task. Although many contractors have experienced the benefits of expanding their market offerings, many more have had unsuccessful experiences causing hardship for the entire organization. Throughout this book, and particularly in Chapter 1, the authors discuss some of the key risks associated with growing a construction company. If the decision to grow the business is consciously made and is consistent with the company's business plan (discussed in Chapter 3), then a complete understanding of the risks involved in this growth activity is necessary. Many contractors deciding to take on a very large project, to expand into unfamiliar locations or to establish a new type of construction, may not see such decisions as dangerously risky. But remember that growth events or decisions preceded the failure of a large number of contractors, and that there is an inherent danger in these activities. With proper planning and controls, the risks can be identified and reduced, but not completely avoided.

Market opportunities can present themselves in a variety of ways: a customer could ask your company to expand geographically, an employee might offer up specific skills, or a new market in need of service can impose a strong pull. Are you considering adding a new line of work or a new market for your existing line of work? The framework presented in this chapter is designed to help.

Growth through market entry provides the opportunity to develop and motivate employees, expand the brand, better serve customers, and mitigate market risks. Unfortunately, statistics show only one out of five market entry decisions are successful. One of the reasons is that most leaders make these types of market entry decisions once or twice in their careers, which are not enough to build experience to identify all the major risks.

Although a standardized decision-making process greatly improves the likelihood of a successful market entry, our survey indicates only 6% of contractors have a formal process in place. This low percentage motivated the authors to work with our collaborators and a large number of contractors in an effort to develop a step-by-step framework that organizes the market entry decision process. It highlights the most important aspects of such decisions, in order to increase your decision's likelihood of success.

DOI: 10.1201/9781003229599-12

Decision Support Framework

The framework consists of ten steps that are organized in three phases as shown in Figure 9.1.

The first phase, definition, clarifies the decision and its context. The second phase, analysis, encourages the decision makers to review similar experiences, focus on the most important decisions factors, and address decision-making issues that commonly hinder success. The third phase, planning, ensures that proper structures are in place to successfully carry out the market entry decision.

The authors and their team developed a step-by-step framework and structured it to ensure the most crucial elements of market entry decision-making, which are critical to the success or failure. The resulting ten steps are organized in three-phases: definition, analysis, and planning.

Phase 1, definition, consists of three steps: (1) understand today's company, (2) define the decision, and (3) choose decision makers and advisors. Decisions that start with a clear purpose are more likely to be successful than those that do not. This phase addresses the definition of the decision, the fit within the company, and who will be involved.

Phase 2, analysis, contains steps 4 through 8: (4) review experiences of others, (5) review lessons learned, (6) assess key factors, (7) identify challenges, and (8) determine the exit strategy. Steps 4 and 5 encourage decision makers to consider a number of similar situations to improve forecasting and creativity when developing alternatives. Steps 6 through 8 focus on the details of the decision at hand. The assessment of decision factors in Step 6 provides tools to assess the essential eight factors as determined through the prioritization workshops. Steps 7 and 8 address specific issues of group decision-making that need to be addressed before moving forward.

Phase 3, planning, concludes the process with Steps 9 and 10: (9) define action items and timeline, and (10) create implementation strategy. These steps are best practices to start developing a plan to enter a new market.

Also included in the framework are four "go/no go" decision points. At these points, contractors are encouraged to reflect on the assessment process. Choosing "go" means the contractor believes it is worthwhile to continue analyzing the market. Choosing "no-go" means critical issues have come up making it impractical to continue. Pausing periodically to reflect at logical points in the process is intended to encourage a thoughtful approach to making a market entry decision. These phases, steps, and "go/no-go" points are shown in the flow-chart in Figure 9.1. The remainder of this chapter will summarize how each step is incorporated into the overall decision support framework.

The time required to enter a new market will vary based on the circumstances. You may have the luxury of time or there may be reasons to act quickly. Regardless, a commitment to this approach and a disciplined review will ensure you have considered the most important factors and are making an informed decision. Contractors who spend more time considering their entry are significantly more

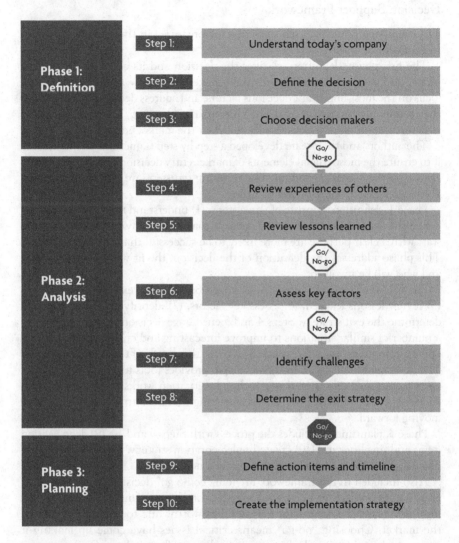

Figure 9.1 Framework to guide a contractor entering a new construction market.

likely to succeed. Decisions are similar to projects; more upfront consideration and understanding improves performance.

Understand Your Company and the Opportunity

First, in order to identify which market opportunities are right for your company, you need to understand your current position to develop alignment among decision makers and provide explanation and justification for the decision. Start by outlining how you got to this decision in the first place. Moreover, data from contractors show a major predictor of market entry success is unwavering commitment from the top company leaders. An influential leader will be necessary to keep the group in alignment and moving forward. The initial steps involve clearly defining the opportunity and articulating your reason for considering market entry, how the new market fits into the bigger picture for your company, and identifying the group of individuals who will ultimately make the market entry decision.

Generally, entering new markets takes the form of either doing the same type of work in a new context (e.g., new geographical location, new type or size of projects) or performing a new type of work (e.g., new trade or technology) in the same context. This is illustrated in Figure 9.2.

Contractors with a defined decision and a clear statement about their reasons for market entry have an average success rating of 4 out of 5. Contractors who do not clearly define their decision and motivation dropped to an average success rating of 2.3 out of 5. We recommend that you take the time to articulate your reason for considering market entry and raise your success rating.

Those who ultimately make the decision may be different for each market entry situation; however, the biggest determinant of success is the commitment and support from the top leaders of the company. Ask yourself the following question: Who will ultimately make this market entry decision? Is it more than one individual? Can company leaders stand behind these decision makers with unwavering support? Who will serve as advisors to the decision-making process? This list might include knowledgeable employees, past/present/potential customers,

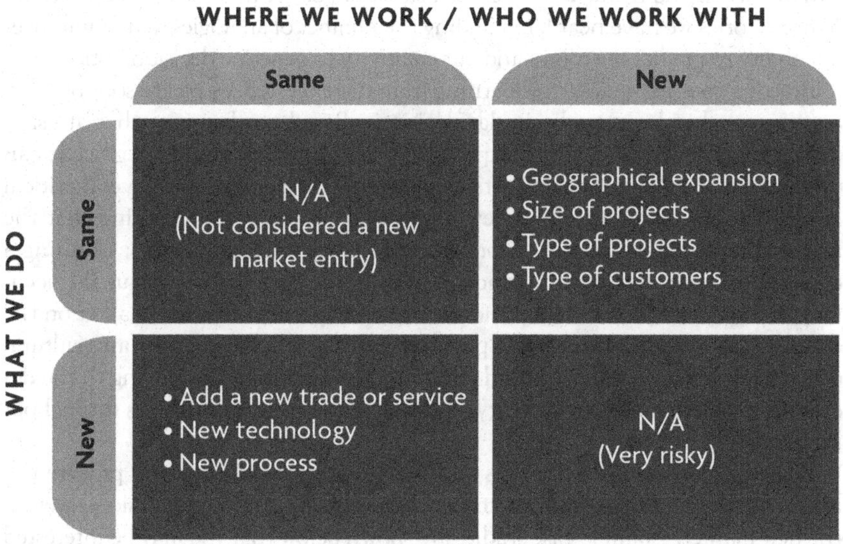

WHERE WE WORK / WHO WE WORK WITH

WHAT WE DO		Same	New
	Same	N/A (Not considered a new market entry)	• Geographical expansion • Size of projects • Type of projects • Type of customers
	New	• Add a new trade or service • New technology • New process	N/A (Very risky)

Figure 9.2 Examples of market entry.

Who is typically involved in advising your company's market entry decisions?

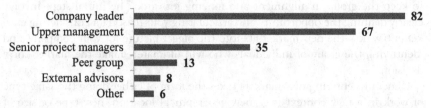

Figure 9.3 Industry survey asking about who is involved in making market entry decisions.

vendors and suppliers, bankers, insurers, accountants, experts in labor market (e.g., internal staff, external network), expert business consultants (e.g., market analyst, acquisition specialist), and so on (Figure 9.3).

After considering your company's mission and vision, formally defining the decision, and choosing the decision makers and advisors, consider whether it makes sense to continue pursuing the market entry decision. After these initial steps, you and your fellow decision makers need to make the first of four go/no-go decisions to enter the new market. If you decide to keep moving forward at this time, it is time to start learning from experiences of others to make sure you avoid their mistakes.

Broaden Your Perspective

Strategic decisions are often made using analogies. In order to make the decision at hand, we should think about a similar situation, either from our own experience or from stories we have heard. Expanding the number of analogies we use improves our ability to predict outcomes and our creativity to generate decision options.

Broaden your perspective by learning from the experiences and lessons of other contractors that have made similar decisions. Based on decisions that are similar to yours, consider realistic expectations for your new market, e.g. what can you expect for profits, start-up costs, labor availability, and so on. Several critical lessons can be learned from reviewing previous experiences: ensuring that the new market is being accounted for separately than existing markets; planning a conservative start; recognizing uncertainties in the new market versus the accuracy of your profit projections; anticipating the impacts of the new market on the company as a whole; identifying potential labor risk; planning for any cultural differences; remaining flexible and continuously adapting; committing to the decision; considering barriers to entry; and maintaining clear objectives throughout the process.

Reviewing previous experiences of others and lessons learned will prepare you and your fellow decision-makers to make the second go/no-go decision to enter the new market. Example case studies are shown below that you may be interested in reviewing.

Mini Case Study 1: Following the Customer

A highly successful completed job prompted a project owner to approach a contractor about working on another project outside of their home state. Previously, the contractor's leadership had discussed expanding into the neighboring state and this offer gave them the opportunity to seriously consider doing that.

A trusted project manager within the company showed an interest in taking on the challenge of leading projects in the new region. It was understood that operations in the new market would have less oversight and support from the home office than traditional projects, simply due to distance. The leaders of the potential project would have to make the project their own, and this interested project manager was willing to take on the challenge. In addition to who would lead the project, the president and the CEO discussed several other factors when considering the new market. They looked at potential profits, support from the bonding company and the bank, the cultural and personality fit, the construction community's response, availability of labor in the area, as well as availability and talent of potential subcontractors. Major suppliers would not be an issue, as the firm planned to work with manufacturers that they have always worked with. After brainstorming all the potential factors the company leaders could come up with, the viability of the move seemed to hold up. After discussions with the potential new leader, the decision was made to expand geographically and accept the owner's request to take the project outside of their typical geographical bounds.

When they expanded the operation was highly project-focused. As customers in the area noticed the productivity, safety, and quality coming from their crew and began requesting the contractor for their own jobs, the project-focused group of employees transformed into a highly profitable branch office. The branch office continued to build a reputation and expand until the 2009 recession took its toll, leveling off work for two years before the company closed down the branch office.

Looking back, the economic struggle of the entire country was not a foreseeable event and the contractor would make the same decision again. The branch office never lost money and, for a while, was a quite profitable arm of the company. The decision to expand and the later decision to pull back were each considered a success.

Mini Case Study 2: Continual Expansion

Since its simple start as a small family business, this contractor continued to grow and thrive, offering valuable construction solutions to an ever-enlarging number of customers. Originally, the metropolitan area in which they were located was growing dramatically, and with it grew the demand for construction. As it sometimes does in the construction industry, development in the area came to a halt, and management at the firm had to discover different avenues to sustain the growth of the company. Expansion to other geographical areas is one of many growth solutions that management was continually evaluating. Several times in the past, the organization has opened successful branch offices in new geographical locations, and it still continues to look for more opportunities to expand in this way.

A nearby location, just across the state line, provided them with a geographical expansion opportunity that was worth further consideration. Managers were looking for an area that exhibited strong union affiliation, was familiar or had high growth potential, and was accessible to leaders traveling in from company headquarters. Because of its proximity to the company headquarters, the nearby location was familiar and easy to visit. It also had a strong union presence. In addition to meeting their minimum requirements, the location offered the opportunity to start with small projects in the market sector where they perform best and with little competition.

After careful consideration, the firm's leadership decided to open an office in the nearby location. Management found an affordable location and a competent leader – a knowledgeable, hard-working manager with many years of experience in the area. Planning had been going on for one year when the new location was opened.

After about five years in business, the branch office shut down. Unforeseen challenges arose with the labor and leadership at the new office. Although the manager running the new office was very experienced and knowledgeable, he ultimately was not the right fit. The new office required a more flexible, entrepreneurial person with a talent for sales and marketing. Labor also proved to be more of a challenge than expected, with three local unions governing the targeted geographical market. The locals were apprehensive about working with an unfamiliar contractor and were very strict with their enforcement of jurisdiction, greatly complicating labor planning.

The contractor believed that having a local address would help combat some of the nervousness that customers and locals have about working with an unfamiliar contractor. In this case, the local presence was not enough. If they were to try growth into this area again, they would consider perhaps expanding by acquiring an established area contractor, to build on the strength of existing relationships and area knowledge.

Mini Case Study 3: Targeted Market

The decision to add another trade at this construction company was provoked by a suggestion from an interested customer. The appeal of the suggestion laid in the ability to create project opportunities in the light commercial market, which required the new capabilities. Together, the president, vice-president, and senior project managers discussed the decision.

At that time, the company was less analytic than it is today. However, factors such as start-up costs, equipment costs, and one- and five-year income projections and cash flow projections were considered as part of the decision process. Most of the market information came from the potential leader of the new department, a trusted friend who worked well with them in the past, and who had worked in the industry for many years.

Two alternatives were considered: (1) adding the new work as a new trade to the company by hiring labor and buying equipment, or (2) subcontracting the new work for light commercial project opportunities. After approximately three months of analysis and deliberation, the decision was made to introduce the new component to the company. The biggest worry from management was that contractors in their current markets would see this as a competitive move. They assured their regular working partners that the move was solely to enter the light commercial market and the company would not be impacted in their traditional projects.

The decision was made to bring in the new work but the additional trade did not last. In less than two years, they decided to pull out of the new market. Looking back, the new department was not well supported by sales and marketing, which potentially occurred out of fear to look aggressive in the eyes of the other contractors they worked with. Also, the company did not fully understand the challenges in the new market, especially perpetual payment delays. If faced with the same decision today, the company would likely conduct a more thorough market analysis for itself, not relying solely on the future champion of the department. The management group would have benefited greatly from learning about the new trade for themselves and likely would not have chosen to enter the new market.

Review Lessons Learned

Reviewing case studies and listening to stories of contractors who have tried to enter a new market can help broaden one's perspective before entertaining the idea of entering a new market. In their research the authors learned that market entry is full of challenges, and they were able to synthesize specific important lessons. Consider the following list of lessons that contractors learned while entering new markets.

Accounting: Many contractors faced difficulties with the accounting aspects once they entered the new market. Be sure you can separate the new market from your other areas of work so you can clearly identify sales, costs, and overhead to ensure your forecasts are accurate. Will your new customers have different invoicing expectations? Can your accounting system invoice fast enough and in enough detail for the customers in the new market?

Conservative start: Those who were successful in entering a new market noted that starting small (low dollar contracts and one project at a time, until the project is fully completed) lessens the inevitable early learning curve. However, dipping a toe into a new market can cause customers to question your commitment. Consider both sides and try to find the right balance for your situation. Start as small as possible to minimize early mistakes or mishaps, but not too small to limit your ability to capture loyal clients.

Ability to plan: Some markets are more stable and easier to plan for than others. Recognize the uncertainty in the new market and the accuracy of your profit projections. An appropriate contingency budget is critical to cope with unexpected challenges.

Existing markets: Entry into a new market always impacts your existing markets. Anticipating the impacts the new market has on the company as a whole can help you recognize ways to not only gain a revenue stream from the new market, but also enhance sales in your current markets.

Labor issues: When expanding geographically or adding a trade, it is important to get a feel for the local market and how it treats unfamiliar faces. This risk should be identified in the decision process.

Cultural differences: Combining groups of people can cause anxiety and pushback. Uncertainty can impact people in unforeseeable ways, especially when their livelihood is concerned. Getting through cultural differences takes a strong commitment from company leaders.

Learn from experiences and adapt: Expect changes and be flexible. The commitment to learn from challenges helped many contractors enhance their new market offerings and gain a competitive advantage after initial struggles.

Commitment is key: Commitment to your decision to enter, or not to enter a market, is a major indicator of success. In the case of a market entry, commitment is necessary at all levels of the company.

Barriers to entry: The barriers you are overcoming to enter this new market will not hold forever. There are likely lessons you can learn today that will help you extend your competitive advantage in this market or successfully enter a new market in the future.

Have clear objectives: Knowing your motivation for entry is a proven key to success. Aligning your reasons for entering the market with how you are measuring success helps everyone clearly see whether goals are being met. When everyone understands the goals, they can adjust their actions to enhance success.

The lessons provided can help the contractor highlight additional risks and inform his or her decision-making process. The next step is critical and consists of assessing the key factors for the market entry decision at hand.

Assess Key Factors

The following eight decision factors were determined to be highly important for success in all types of market entry, and all eight factors need to be assessed adequately.

1   Strategic/cultural fit of the new market with the company's overarching strategies, competencies, and culture;
2   Experience, abilities, and drive of the person who will champion the new market;
3   Market need, including current size and potential growth/shrinking of the new market;
4   Competition already existing in the new market;
5   Competitive advantage, edge, or niche the company may possess in the new market;
6   Start-up costs needed to enter the new market;

7 Year-to-year profit projections of new market revenues versus costs; and
8 Investment capital readily available to commit to the new market entry.

An in-depth assessment of these eight factors serves as a necessary foundation for your decision to enter the new market. For example, Factor #3 around market need can be gauged in many ways, such as using published market data or studying proposal lists to identify how much of the work is new construction, whether pre-qualification is required, any noticeable geographic trends, which clients produce the most projects, and so on. Another example, Factor #4 around competitive advantage can be assessed using a SWOT analysis, which stands for Strengths, Weaknesses, Opportunities, and Threats. It is a tool to build awareness around these four elements of your new market environment, both internal and external to the company.

Other factors to consider, which may be important in certain situations, include bonding company acceptance, contract requirements, customer acceptance, market knowledge, labor commitment, staff availability, standards and regulations, and support from the organization.

Plan for the Worst

If your decision is to move forward with the market entry, that is exciting news, but you still want to plan for the worst. Identify potential challenges using a *what-if session* to create a safe space to voice reservations about the decision. Ask all decision makers and advisors to meet and *assume* as though everything in your new market has gone wrong. Allow two minutes of silent time for each person to individually write down all the reasons the market entry could have gone poorly. Then, go around the room and allow each person to read their reasons for failure. Based on these reasons and the above steps, you can now determine the exit strategy.

Once a new market is entered it is a difficult decision to exit, so the time to set the limits of the effort is before you start. Document (a) how much you will invest, (b) how long you will continue if the new market is not developing as expected, (c) what parameters will trigger an exit, and (d) how you will implement a market exit by outlining the steps and actions that will be taken in case this unfortunate scenario unfolds.

Cash flow (not to be confused with profitability) is often a critical consideration at this time. For new ventures or expansions in the construction industry, cash flow will be negative for a period of time due to initial funding or borrowing needs to cover the upfront startup costs. It is common for cash flow to be negative for the first two years, even for a profitable effort. Often cash flow is negative for three years. In your exit strategy, you may want to consider deciding in advance to withdraw if cash flow is not positive in four years. After identifying the potential challenges and determining your exit strategy, you will make the fourth and final go/no-go decision to enter the new market.

Go versus No-go versus Delay

The framework presented in this chapter is designed to help you make a decision to enter (or not) a new market. The final answer can be go, no-go, or delay the decision for later. If you have made it through all the steps and you decided to go ahead and enter the new market, you will need to start planning for the market entry. Start by defining key action items needed to reach your goals, along with a reasonable timeline for each, and assign responsibilities accordingly.

As you move forward, make a commitment to follow up on each of your action items and ensure your adherence to the goals and original reasons for entering the market as identified in the beginning of this chapter.

By using this framework to structure your market entry decision, you have leveraged some of the industry's key knowledge and experience to increase your decision's probability of success. However, market entry decisions are inherently risky, as discussed throughout this book. The contractor needs to be sure of this decision, the risks it entails, the price of growth, and the limits of this venture. With proper planning, the risks can be identified and reduced, but not completely avoided.

Review Questions. Check All That Apply.

1   How can market opportunities present themselves?
    a   Customers ask company to expand geographically
    b   Employee has specific skills
    c   New market in need of service
    d   All of the above

2   What does an industry survey define as the top advisor about market entry decisions?
    a   Upper management
    b   Company leader
    c   Peer group
    d   None of the above

3   How long do the authors say it is common for cash flow to be negative for a new venture?
    a   6 months
    b   1 years
    c   2 years
    d   3 years

4   What do the authors say is a proven key to success?
    a   Understanding the new market
    b   Knowing your motivation
    c   Agreement of lenders
    d   Available personnel

5   What do the authors say the contractor need to be sure of in the decision to enter a new market?
    a   The risk it entails
    b   The price of growth
    c   The limits of this venture
    d   All of the above

Critical Thinking and Discussion Questions

1   Explain the ten steps for a successful market entry.
2   Discuss the meaning of Figure 9.2. Give examples of market entry.
3   Outline specifically who you would rely on for advice on a market decision.
4   Discuss the eight key decision factors and list those you consider most important.
5   Explain what the section on *planning for the worst* means to you.

10 Project Selection

There are no bad projects – just bad matches of contractors to projects. Project selection is one of the most critical decisions construction enterprises make. The risk associated with project selection can be measured in advance and is directly associated with the construction organization's past experience with similar work.

The performance of an operating entity, such as a manufacturing plant, improves with repetition. In contrast, construction companies do not usually have enough repetition from project to project to experience much improvement. In this industry, almost all new projects differ from previous projects in varying and measurable degrees. The measurement of risk depends on the project's characteristics, experience of the team members involved, and other project risk factors.

Project Risk Impact Factors

Key factors that impact project risks include (1) project size, (2) project type, (3) project location, (4) project teams, and (5) complexity of design features (Figure 10.1). These aspects can be measured and weighed to produce a numeric measurement of risk potential and can also be tracked to measure project performance. The more experience a contractor has with similar projects, the greater the likelihood of a successful estimate, production, and completion at a profit. Experience with fewer similar projects translates to lower level of success probability and presents greater risk. Let's explore these factors further.

1 Project Size

Construction organizations produce projects of varying sizes. Most organizations have a good number of small projects (small is relative to their organization's average project), a reasonable number of mid-size or "average" size projects, and a lesser number of large projects.

Typically, small projects may be performed as a service for regular and valued clients and usually lead to high profit as a percentage of sales. However, in total the profit is not large enough to support the organization. Some contractors refer to small projects as nuisance work, while others say small jobs help pay the rent.

DOI: 10.1201/9781003229599-13

Midsize "average" projects generally earn reasonable profits as a percentage of sales, but this percentage is smaller than that of small jobs. For most construction enterprises, midsize projects are the projects the company thrives on. They can be described as the company's regular work or typical projects that most of their estimators can price, most superintendents and project managers can build, and these projects often perform as expected. Because of the depth of experience with these size projects, the level of confidence proposing on them is high and the anticipation of successful production and completion is equally high.

A construction company will usually have fewer large projects, which earn less than the midsize projects in terms of profit as a percentage of sales. These large projects help contractors reach critical mass, support growth appetite, and support the interests of key employees. These projects differ from midsize projects because (1) not every estimator can price them, (2) top management takes an interest in the estimate and pricing of the work, and (3) the day before the estimate is submitted may be a long day because less experience demands caution. The greater concern about proposing on a large project (compared to a mid-size project) indicates the greater amount of risk. However, it is important to note that these concerns are typically caused by the organization's limited experience with these larger projects, relative to the smaller and midsize jobs they have successfully completed in the past. The amount of similar past experience with any project is directly proportional to the level of confidence in pricing and anticipation of success. Experience with fewer large projects translates to a lower level of confidence (whether that is recognized by the organization or not) and presents greater risk.

If a small project does not perform well, or loses money, there is typically not enough money involved to significantly impact the company financially. A midsize project that loses may concern the organization, but there are usually enough other successful midsize projects to pick up the slack. In contrast, a large project that underperforms or loses money has the potential to have a significant impact on the company's financial condition.

2   Project Type

Similar to project size, prior experience with the type of project is directly proportional to the likelihood of successfully proposing and delivering a project profitably, on time, and on budget. For example, if an organization has been successful constructing relatively straightforward warehouses and strip shopping centers, and then decides to attempt their first complex sewage treatment plant, they would have no institutional experience on this type of project. The likelihood of successfully pricing and producing the work would be limited, resulting in high risk compared to an organization that regularly builds treatment plants. The amount of risk will be different for each construction organization perusing the work, depending on the institutional and individual experience of each enterprise. If many of their prior projects are of similar types, there is less impact on risk.

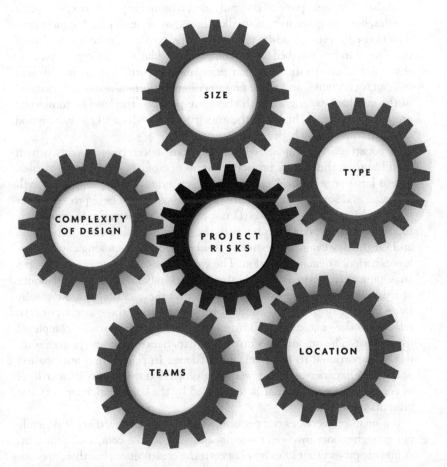

Figure 10.1 Key factors that impact project risks.

3   *Project Location*

Because construction work is produced slightly differently across various parts of the country, experience working in a certain geographic area impacts the contractor's likelihood of success and risk. For example, if a contractor is attempting their first project in an adjoining state where the contractor has not built in the past, the project would obviously be outside the organization's experience and may involve a challenging learning curve. The likelihood of success would be less than if the project were in the contractor's current state, and a fundamental risk to profitability will exist.

Similar to project type, departure from the geographic area creates greater risk. Contractors will be required to resolve potential differences that may arise due to shifting from their traditional areas to new geographic areas. These differences include (1) labor issues and skill levels, (2) subcontractor availability, pricing, and expectations, and (3) other local customs that may impact how the work is managed or performed.

4 **Project Teams**

Experience is accumulated institutionally but is captured individually. The number of members on a project team with direct and relevant experience improves a project's delivery performance. This is evident in new qualifications-based selection (QBS) processes specifically evaluating the key team members' experiences and resumes as a critical weighted component of the firm's submitted statement of qualifications (SOQ) when proposing on a project. Considerable institutional experience does not imply that all individuals within the organization have that experience as well. If an organization has experience with similar project size or project type, but no project team member assigned to the work has personal experience in similar project size or type, then the project team has little experience related to the project in question.

If a team is composed of individuals that have experience on similar projects, the project risks are significantly reduced. If the project team also has positive experience working together on projects in the past, the risk is reduced even further. Experience of the team members and experience working together impact the amount of project risk.

5 **Complexity of Design**

Most buildings in the US are rectangular or simple in shape; the largest portion of roads and highways are straight for much of their length; and bridges and tunnels are typically straight to the extent possible. This is not to suggest there are no curved buildings, roads, bridges, or tunnels, but simply that there are fewer of them. Therefore, the lion's share of experience collected by construction enterprises is with familiar and generally conforming shapes and designs. Each construction organization has its own institutional and individual experience, but it is safe to say that the collective experience of the construction industry is with conventional projects following established designs.

Therefore, if a project has curved walls, windows, roofs, or unique elements, then it is outside the direct experience of most organizations. The same goes for out-of-the-ordinary roads, highways, bridges, industrial projects, one-off projects, or projects with matchless features. These projects can and will be built; however, there is considerable project risk in pricing and producing them because they fall outside the realm of commonplace or conventional projects. Recently, a small number of construction companies have specialized and become known for tackling complex unusual projects, based on a growing portfolio of complex project experiences.

Measuring Risk of Project Success

A breakthrough in the authors' research of measuring the potential for project success in advance is the realization that it is not only about the project's built-in inherent risk as a stand-alone project. A large portion of the risk to project success is a measurement relative to the selected contractor.

For example, a contractor might say that a totally unique, one-of-a-kind project is risky in and of itself. However, this is not always the case. A contractor that has the closest experience to such a project would still have risk, but their risk is smaller than another contractor with less relevant experience. The less related experience the contractor has, the more risk they would be taking and the more likelihood that project becomes a losing project.

Avoiding Losing Projects

The construction business growth model of the past does not always work, especially in a cyclical industry. Many of the construction enterprises the authors are familiar with embrace or are strongly influenced by a business model driven by growth. This type of model puts too much emphasis on increasing sales or maintaining sales, sometimes for the sole purpose of covering overhead.

A common belief in the construction business is "*If you're not growing, you are going backwards.*" However, this and many other well-established construction industry beliefs need to be reconsidered, for example: *growth is always good; overhead is a symbol of success and not to be surrendered unless absolutely forced to; cutting overhead is an admission of failure; downtimes are bad news but a natural part of the industry; the industry is not necessarily cyclical; and unprofitable work is just part of the business.* These beliefs cause many to go after any available projects, in good times or bad times, or outside of their organization's experience. A lack of experience dramatically increases risk, but many contractors still refuse to give up hard-earned growth. A contractor that takes on work for which they have limited experience can find themselves in a situation the authors call the "80/20 syndrome."

80/20 Syndrome

A study of hundreds of failed construction companies confirms that a large majority of these financially distressed firms also had profitable work. However, their problem was that the losing projects consumed the profits from the successful projects. Those successful projects also had to cover the overhead costs that the losing projects failed to contribute to. It is fair to say that many, if not most, construction enterprises have occasional losing projects. As mentioned earlier, it is a well-established industry belief that *some unprofitable work is just part of the business.* The failed companies in the authors' database had on average 80% of profitable projects and 20% of unprofitable projects. In these failure scenarios, the amount lost simply grew beyond what the profitable work could support (Figure 10.2). While profit reduction was a major issue, overhead was often the larger problem. A case study example will further describe this notion.

Case Study: A contractor with $10 million annual sales has $1 million in overhead (the authors are not recommending 10% overhead; this number is used here

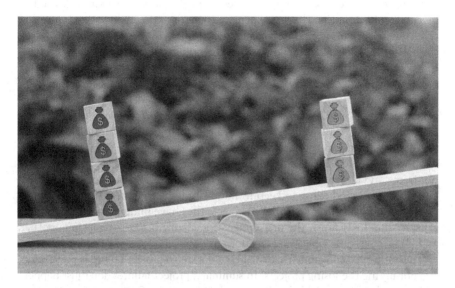

Figure 10.2 The amount lost simply grew beyond what the profitable work could support.

just to simplify the calculations). Eighty percent of the work, or $8 million, is generating an $800,000 contribution to the 10% overhead and $240,000 in net profit (3%). However, the remaining 20% of the work, or $2 million, has a loss of 5% or negative ($100,000). This loss is covered by the $240,000 net profit earned by the 80% profitable work, leaving $140,000. However, the $2 millions of un-profitable work contributed nothing toward the 10% overhead, so $200,000 of needed overhead costs must now be funded but there is only $140,000 remaining from the profitable projects. This results in a ($60,000) loss for the company for the year. In this example, 80% of the organization's work made full profit, which is highly unlikely for a firm experiencing losses on the other 20% of its work. Even with the unlikely high profit there is a company-wide loss for the year when only 20% of its work underperformed. In most cases the loss would be greater when the 80/20 syndrome occurs because in most underperforming organizations that have losing projects, the remaining work often does not make full profit, but some lesser amount. If unprofitable work is just part of the business, then at the very least, minimizing the amount of unprofitable work must be a critical priority of management. The critical question here become the following: How can a com-pany minimize the amount of its unprofitable work? The answer resides in project selection.

Project Experience

Spending decades studying, defining, and measuring the causes of business fail-ures motivated the authors to study how to reverse or manage losing projects while they are still in progress. The research confirmed that it is incredibly difficult to reverse troubled projects, hugely expensive to manage them, difficult to eliminate the damage once done, and impossible to minimize costs already incurred. What the authors did discover instead was that the primary cause of project failures was inexperience with the type of work, project, or process.

Experience is a *"Paradox of Industrial Proportion"* because it is fair to ask, *"If experience is critical to success, how do I gain the initial experience?"* This leads to another question: *"Should a construction enterprise ever take on work they have never performed before?"* This issue is a matter of risk tolerance and timing, and a decision about emphasizing profit or volume – a contractor's choice to make. However, contractors should understand that knowledge and experience are purchased, are not free, can be extremely expensive, and there are serious risks associated with the process. A good thing to keep in mind would be to expect the unexpected. Organizations that attempt work on projects they have limited experience should (1) start small, (2) finish the first project completely before attempting another, and (3) attempt only what you can afford to lose.

The authors would argue that the best solution to project failure is prevention, i.e., not taking the losing projects. In summary, project failures don't just happen. They are jobs that are deliberately pursued and captured. Therefore, as mentioned earlier, the answer resides in project selection.

Project Selection Program

A systematic process can be used when a contractor is deciding to go after a given project or not. The five project risk impact factors discussed in the earlier sections can be weighted and placed in a numeric formula that provides a measure of project risk. This section presents a tool that will greatly enhance project owners', contractors', and subcontractors' project selection process. The tool supports the contractor in making better-informed decisions, capturing profitable projects, and screening out losing projects.

The need for a process to screen-out losing projects at the pre-project stage in order to enhance profitability led to the development of the project selection program. During the testing phase of the program, numerous contractors ran the program on three of their completed projects, which included two successful projects and one project they wish they had not taken. Respondents answered the 26 individually weighed questions, which gave them a numerical score that in fact accurately measured how the firm's prior experience lined up with the project's specific requirements. The results were consistent, and the contractors confirmed the tool is a predictor of the potential success of their tested projects. Potential users of the program can do the same, experimenting with three of their own completed projects to gain confidence in its use. Figures 10.3–10.5 show screenshots of the Excel-based program.

The "Project Selection Program" is available at no cost on this book's companion website: www.routledge.com/9781032134734

The tool is located under *Support Material* toward the bottom of the page. Note that this tool works best on MS Windows machines.

Project Selection Program v1.1 © Thomas C Schleifer, Ph.D. tschleifer@q.com

	50.0%

Project Name:

Completed By: Date:

Introduction: Click here

Directions: Click here

Change weights: Click here

Click on question for more detail

Section 1 - Project Fit	Weight	*Default Weight*
1.1 Has the firm successfully completed projects this size (this large)? ○ Never ○ Few ○ Often	10 pts	*10 pts*
1.2 Has the firm successfully completed projects of this type? ○ Never ○ Few ○ Often	10 pts	*10 pts*
1.3 Has the firm successfully completed projects in this geographic area? ○ Never ○ Few ○ Often	10 pts	*10 pts*
1.4 Has firm successfully completed projects with this Owner/GC/CM? ○ Never ○ Few ○ Often	10 pts	*10 pts*

Section 2 - Firm's Experience	Weight	*Default Weight*
2.1 Has the firm successfully completed projects with this Owner/GC/CM's representatives, inspectors, etc.? ○ Never ○ Few ○ Often	2 pts	*2 pts*
2.2 Has the firm successfully completed projects with this designer/architect/engineer? ○ Never ○ Few ○ Often	5 pts	*5 pts*
2.3 Has the firm successfully completed projects with this designer/architect/engineer's field personnel, inspectors, etc.? ○ Never ○ Few ○ Often	2 pts	*2 pts*
2.4 Has the firm's estimating team priced this type of project? ○ Never ○ Few ○ Often	4 pts	*4 pts*
2.5 Has the firm's anticipated field team (PM, Supt, Foreman) built this type/size of project? ○ Never ○ Few ○ Often	6 pts	*6 pts*

Section 3 - Project Complexity	Weight	*Default Weight*
3.1 Is the required project schedule reasonable? ○ No ○ Unknown ○ Yes	4 pts	*4 pts*

Figure 10.3 Project selection program (screenshot 1 of 3).

3.2	Is there sufficient labor available to produce the project? (consider available skill levels)	5 pts	5 pts
	○ No ○ Unknown ○ Yes		
3.3	Has the firm had prior successful experience with the intended major subcontractors/vendors/etc.?	5 pts	5 pts
	○ No ○ Unknown ○ Yes		
3.4	Are there unusual access, room-to-work, parking, dust, noise, storage space, traffic, etc. issues?	3 pts	3 pts
	○ Yes ○ Unknown ○ No		
3.5	Are there unusual ground or surface water conditions?	3 pts	3 pts
	○ Yes ○ Unknown ○ No		
3.6	Are there rock, soft soils, unusually deep excavations,etc. that may complicate or delay the project?	3 pts	3 pts
	○ Yes ○ Unknown ○ No		
3.7	Are there unusual, circular, curved or angled design elements: i.e. walls, bridges, roofs, structures or other that may complicate/delay the project?	4 pts	4 pts
	○ Yes ○ Unknown ○ No		
3.8	Are there critical long-lead, sole source or unusual items that may delay the project?	2 pts	2 pts
	○ Yes ○ Unknown ○ No		

Section 4 - Financial and cash flow impacts		**Weight**	*Default Weight*
4.1	Is the project owner's financing in place?	2 pts	2 pts
	○ No ○ Unknown ○ Yes		
4.2	Does the firm have experience with the payment processes of this owner/GC/CM?	2 pts	2 pts
	○ No ○ Unknown ○ Yes		
4.3	Will this project strain the organization's resources: i.e., financial, equipment, manpower, management, or other?	3 pts	3 pts
	○ Yes ○ Unknown ○ No		

Section 5 - Contract issues		**Weight**	*Default Weight*
5.1	Are contract terms reasonable?	1 pts	1 pts
	○ No ○ Unknown ○ Yes		
5.2	Are there any unusual payment provisions, retainage issues or liquidated damages that will affect cash flow?	1 pts	1 pts
	○ Yes ○ Unknown ○ No		

Figure 10.4 Project selection program (screenshot 2 of 3).

5.3 Are there any unusual bonding requirements?			**1 pts**	*1 pts*
○ Yes	○ Unknown	○ No		

5.4 Are there any unusual insurance requirements?			**1 pts**	*1 pts*
○ Yes	○ Unknown	○ No		

5.5 Are there any unusual indemnity issues?			**1 pts**	*1 pts*
○ Yes	○ Unknown	○ No		

Section 6 - Additional Issues to Consider (*these are not scored*)

6.1 Do you believe you have a competitive edge on this project?

Going after a project generally has a cost associated with it along with potential lost opportunity if the use of resources to get the project prevents or impacts going after another job or reduces efforts of attracting other projects. This makes the measurement of a competitive edge or lack of one significant. Where there is a competitive edge on a project it may make sense to invest more resources than on one where there is less or no competitive edge. This is difficult to measure, but for most contractors it is in intuitive.

6.2 Do you believe another contractor(s) has a competitive edge on this project?

Realistically recognizing any competitive edge other contractors may have is valuable knowledge and can assist in a project selection decision. Most are familiar with projects where a favored contactor was the owner/GC/CM's obvious choice and many other contractors wasted a lot of effort and resources only to see the obvious choice selected. Screening out projects that a firm has limited chance of capturing saves considerable resources which can be applied to projects with a higher likelihood of success and can potentially increase sales.

6.3 Is there consensus within your organization about perusing this project?

When everyone in an organization is excited about pursuing a particular project the odds of capturing that work go up through the quantity and quality of the energy expended. The converse is true when there is little enthusiasm for a project making capturing it more difficult. There are situations when not going after a project is the best project selection decision. If some or many are concerned about a certain project it may be the wrong job for the company

6.4 Is your "gut feeling" about this project negative or positive

The expression; "If all else fails trust your gut" can be a valuable inclusion in the project selection process. The information collected from the above questions is of great value, however, if persons in responsible positions within an organization still have a gut feeling about a positive or negative decision it probably makes sense to reevaluate the data before making a final decision.

6.5 How does the project fit with the existing/anticipated backlog?

A contractor whose current work is running smoothly has only limited exposure in taking on additional work. However, a contractor managing a troubled job or two may encounter additional risk in adding the next project. A well-managed and organized backlog has reasonably low exposure and makes additional, prospective projects attractive. A troubled backlog has higher risk which may indicate that this may not be the time to add another project.

6.6 Does the contract include Consequential Damages or are they expressed or implied?

Consequential Damages are damages resulting as a consequence of the project. The owner is the usual party damaged, the public and others may also be included. Fairly common are damages that result when a revenue-generating project is delivered late such as a restaurant, hotel or retail business. The damages include loss of revenue because the intended use is delayed which can result in serious potential liability for a contractor that causes this to happen. Some contractors will not undertake such exposure which unlike liquidated damages cannot be measured in advance.

Figure 10.5 Project selection program (screenshot 3 of 3).

About the Project Selection Program

If you click on *Introduction, Directions,* or *Change Weights* cells, comment boxes appear with the information below.

Introduction: There are no bad projects, just bad matches between projects and contractors. The probability of success in go/no-go decisions in project selection can be accurately measured in advance and are directly associated with the construction organization's experience with similar work. The elements impacting success have been weighted to produce a numeric scale of how well the prospective project matches the firm's experience: designed for general contractors, subcontractors, and construction managers; effective for architects, engineers, and vendors; supports owner's contractor selection.

    Directions: Answer the questions without overthinking them. Total possible scores range from 10% to 90%. The program is based on a firm's knowledge about a potential project. If nothing is known about a project, probability of success is a "toss up," which in statistical language translates to 50/50 or a 50% probability. Therefore, the program score opens at 50% before adding knowledge about the project. As knowledge is added it raises or lowers the score. A higher score represents a higher potential for success and a lower score reduced potential. There is benefit to multiple people within an organization score projects separately or in collaboration.

    Change Weights: The weighting of questions is derived from 40 years of measurement and experience and considered accurate. While it is not recommended, users may change the weight for any question they feel may be more appropriate for their circumstance. The default weights are shown to note how far a change is from the program weight. Weighting is based on statistical analysis which takes into account the likelihood that an event will occur and the severity if it does occur. For example, a costly issue (severity) occurs historically in a small percentage of projects it is weighted lower than an issue with high severity and high occurrence. Users should change weighting values carefully keeping in mind that varying one point of a 10-point question revises it by 10% while adding one point to a one-point question revises it 100%.

Final Thoughts: Impact on Overhead

If you screen out potentially losing projects and decide to not go after them, sales may fluctuate, while profit and efficiency go up and business risks diminish. While overhead should always be carefully managed because construction is a cyclical industry, fluctuating sales make this attention to overhead even more important.

The successful contractor of the future will be profitable in good markets and bad markets alike, by adding some science and knowledge into the management of their business. Unless growth is deliberately part of the company's strategic plan and its associated "cost" is fully acknowledged and planned for, the drive for size and growth will be substituted with a more measured drive for prosperity, which is measured in profitability. This can be done by not taking losing jobs, which is accomplished by prudent project selection and judicious management of overhead. For many, this will be a paradigm shift because the predominant current growth model supports growing overhead.

If a contractor has effective control of their overhead, using practices shared in the earlier chapters such as flexible overhead, they can have the luxury to select only projects they are experienced with. This dramatically reduces the number and severity of losing projects, maximizing profits. Some projects just don't fit a contractor's experience. As a quick reality check, if a contractor is asked, "*Is there one project from any prior year that you wish you hadn't taken?*" and the contractor answers "*yes*," that contractor is in effect saying that his or her preference is to have had a smaller size company that year with more profit. Prudent project selection does not stunt growth; it simply redirects focus from growth to profitability.

Review Questions. Check all that apply.

1 Which of these is not included in the authors' project risk impact factors?
  a Complex design features
  b Project teams
  c Project schedule
  d Project location

2 What do the authors claim about the business model of growth:
  a It is appropriate for construction
  b It puts too much emphasis on increasing sales
  c It is the most profitable model
  d It helps ensure overhead is covered

3 What are differences in shifting to new geographical locations?
  a Labor issues and skill levels
  b Subcontractor availability, pricing, and expectations
  c Local customs about how the work is preformed
  d All of the above

4 The project selection tool supports the constructor:
  a Making better-informed decisions
  b Capturing profitable projects
  c Screening out losing projects
  d All of the above

5   What do the authors say is the best solution to project failure:
   a   Hard work
   b   Careful planning
   c   Prevention
   d   Proper scheduling

Critical Thinking and Discussion Questions

1   Discuss the five key project risk impact factors and rank them by order of importance.
2   Detail why and how "complexity of design" impacts risk.
3   Explain what the authors mean by the 80/20 syndrome.
4   Discuss why project experience is so important in project selection.
5   Explain the use and the claimed advantages of the project selection program.

11 Managing Subcontractors and Equipment

Two of the largest costs, and hence risk exposures, in construction execution in the field are subcontractors and equipment. The management of these two areas is critical to success but is too often left to field supervision to accomplish with limited guidance. Many interviewed subcontractors describe inconsistent treatment, direction, and management between different projects for the same general contractor or construction manager. Many have said they try to find out early, in advance of bidding, who will be running a particular project, and some even alter their price accordingly. In working with hundreds of construction enterprises, the authors found little similarity in equipment ownership, leasing, management processes, policies, or procedures. There are certainly multiple ways to manage equipment well, but there are also cost and efficiency advantages associated with operating consistently throughout an organization. The two sections of this chapter address subcontractor management and equipment management, including recommended policies, processes, and procedures.

Subcontractor Selection

Managing subcontractors and vendors begins with the selection of the right company or supplier for the job at hand. This task is more difficult than it sounds. It is not always possible for contractors to find the time or resources to select the best available firm that is ideal for the job. Challenges that hinder the selection process include project-related schedule constraints, availability of subcontractors in a given area, owner or designer preferences, favorites from top management, price constraints, and so on.

It should go without saying that selecting the wrong subcontractor, for any reason, has the potential for serious consequences. Many losing projects are a result of the non-performance or failure of one subcontractor on the job. Objective subcontractor selection is important, even critical, to project success. It requires choosing the most experienced and qualified subcontractor who can get the job done with minimal risk. Subcontractor selection criteria should include examining past experience with similar project sizes and types, and experience with other contractors, at a minimum.

DOI: 10.1201/9781003229599-14

The greatest potential for project success is when the size of the subcontractor's work is around their average project size based on the past several years. Projects approaching the largest project a subcontractor has ever performed present greater challenge than projects of "average" size for that subcontractor. Subcontractors always have more experience with their midsize work, which is also less of a challenge for them to finance. It is safe to assume that a contractor will find it easier to obtain the necessary workforce to supply their midsize projects. They will more likely have the right equipment available and should be able to accelerate the work if it becomes necessary.

Along the same lines, the right subcontractor will have considerable experience in the exact type of work involved and will be able to demonstrate that to your satisfaction. This will be challenging if you are working in an unfamiliar area with unfamiliar subcontractors. Size and type of construction impact a subcontractor's performance and ability to meet the schedule, so the closer the project is to their average size and type of work, the more likely they will recognize and achieve schedule requirements. It is business-as-usual when the project is similar to what they perform regularly.

Another effective way to check the capabilities of a potential subcontractor is through exploring their prior project performance with other contractors in the area. Ask if they pay their suppliers on time, if they meet schedules, if they cooperate with other trades on the job, and so on.

Subcontractor Management

After a subcontractor has been selected, they must be managed effectively. It is critical to inform subcontractors about your expectations because if they know what is expected of them before the work commences, they are better prepared to meet the expectations for the project. Setting expectations early allows you to deal with any issues and objections upfront.

The authors advise construction companies to develop a well-thought-out written subcontractor management policy, institute it company-wide, and provide it to bidders during the preconstruction or pre-bid phase. A subcontractor's efficiency and productivity on a project are directly proportional to how well they are managed. The practice of developing a subcontractor management policy allows a contractor's field supervision team to lead and coordinate appropriately and consistently.

The authors provide a sample subcontractor management policy in this chapter, for a theoretical company called OK Construction, Inc. This sample is only an example and can be expanded or reduced to suit different contractors' individual purposes.

The bedrock of subcontractor management is to recognize and expect that subcontractors will make a profit, and let them know you care about that. The first section of the subcontractor management policy is the introduction, which is beneficial to both the contractor and subcontractors, and is informative for designers, owners, and others. This first section should be distributed to the aforementioned parties and explained to them (Figure 11.1). This is not a confidential policy and depends on transparency for all the parties concerned.

The second part of the policy is a component of the contractor's corporate policies, such as an employee handbook, and is therefore designed for internal use only (Figure 11.2). It is not confidential per se, but is not intended for mass distribution either. This part of the policy explains and teaches a contractor's supervisors and managers what the firm expects and "how to" manage their subcontractors. It particularly addresses the importance of equal treatment of subcontractors and the necessary rigorous coordination of all involved stakeholders.

Construction projects rely on collective efforts of general contractors, construction managers, subcontractors, vendors, designers, engineers, owners, and many more stakeholders. Subcontractors are a key cog in this process, especially since they often deliver a large portion of the scope. If they are not valued as key members of the project team and the projects does not run smoothly, everyone suffers.

Effective contractor and subcontractor selection and management can be difficult and complicated, but it can make the difference between a successful project or a failed project. It is critical to identify and select the most qualified contractors for each project and to administer clear management policies. Now that we have discussed subcontractor management, the next section of this chapter presents another important business management element for the successful and prosperous construction enterprise: managing equipment.

Equipment Management

Accounting for a construction company's equipment has caused serious financial problems for many contractors. The process of evaluating and accounting for all costs associated with equipment ownership and usage is complicated by the fact that some equipment costs do not appear on everyday invoices; therefore, they are not incurred in the course of operations. These costs, while not easily recognized, will eventually have to be paid. Although various formulas are used to cope with managing a contractor's equipment costs, the total cost of ownership is often misunderstood. All costs, including those hidden, must be recognized and planned for in advance.

Contractors face many equipment concerns and decisions, which include whether to buy or lease equipment, which equipment to invest in, and when to invest. Equipment-intensive contractors, such as road builders, landscapers, and sheet metal contractors, have to make these equipment-related decisions often, and the decisions have a long-term and profound effect on the organization's success. Contractors whose work is not equipment-intensive may still have similar equipment-related concerns and decisions to make, but less effort might be put into these decisions since equipment is not a significant part of their business. For instance, debt service and maintenance costs still exist for contractors who own very little equipment, but these costs exist in a much larger way for equipment-intensive contractors.

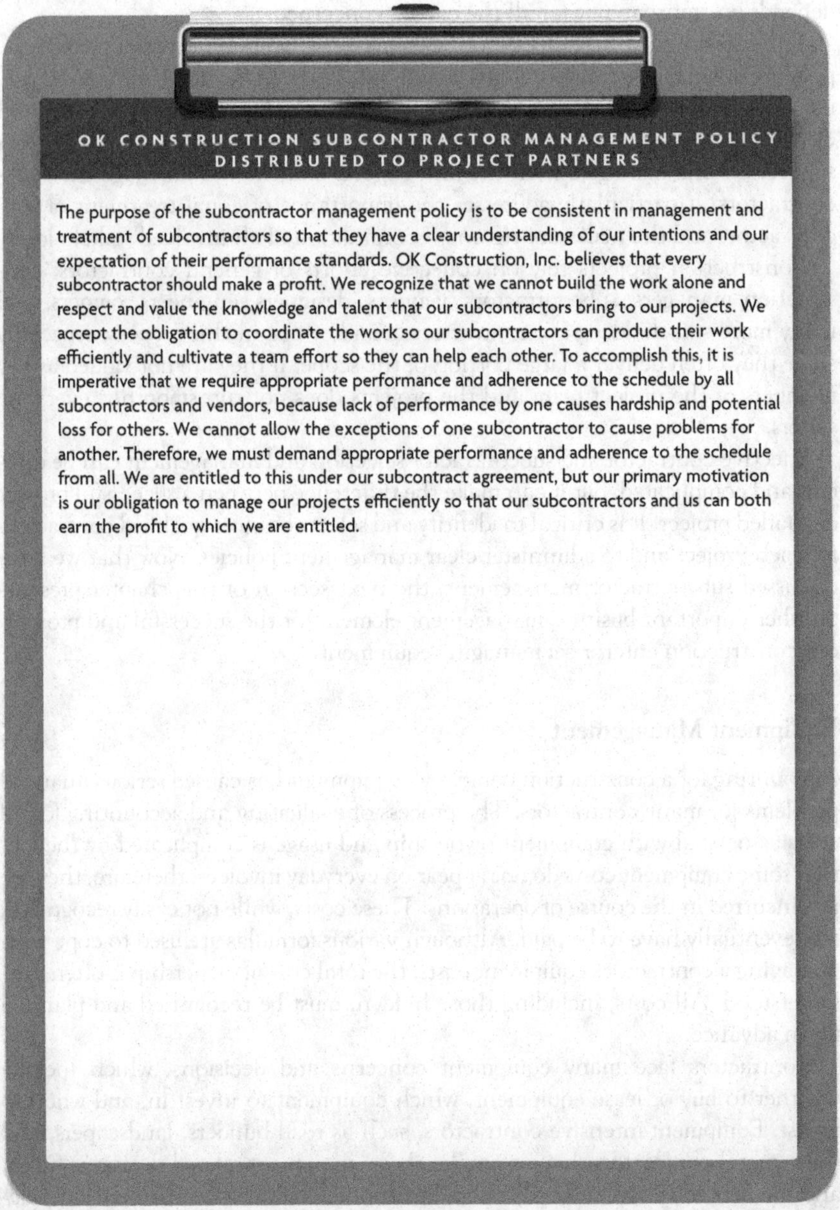

**OK CONSTRUCTION SUBCONTRACTOR MANAGEMENT POLICY
DISTRIBUTED TO PROJECT PARTNERS**

The purpose of the subcontractor management policy is to be consistent in management and treatment of subcontractors so that they have a clear understanding of our intentions and our expectation of their performance standards. OK Construction, Inc. believes that every subcontractor should make a profit. We recognize that we cannot build the work alone and respect and value the knowledge and talent that our subcontractors bring to our projects. We accept the obligation to coordinate the work so our subcontractors can produce their work efficiently and cultivate a team effort so they can help each other. To accomplish this, it is imperative that we require appropriate performance and adherence to the schedule by all subcontractors and vendors, because lack of performance by one causes hardship and potential loss for others. We cannot allow the exceptions of one subcontractor to cause problems for another. Therefore, we must demand appropriate performance and adherence to the schedule from all. We are entitled to this under our subcontract agreement, but our primary motivation is our obligation to manage our projects efficiently so that our subcontractors and us can both earn the profit to which we are entitled.

Figure 11.1 Sample subcontractor management policy for external use.

OK CONSTRUCTION SUBCONTRACTOR MANAGEMENT POLICY
FOR INTERNAL USE

It is our obligation to be certain that subcontractor foremen understand, from their first day on our project, what is expected of them, that they report to the OK Construction Superintendent, and are instructed to comply with their directions. If they do not fully understand this, the Superintendent is to contact the subcontractor's office and/or the OK Construction Project Manager and get the issue resolved within the first week they are on site. As professionals and in a professional manner, the Superintendent must set the record straight at the beginning of the project so that they can require performance throughout the project. We will not accept "I can't get done this week because of this, this, and that," because 95% of the "reasons" can be overcome with hard work and/or overtime. We are entitled to take the position: "It's not our problem; it's yours." Further, if other foremen see nonperformance being accepted, they relax and an air of "everything can be worked out" takes over. We will not accept nonperformance or inconsistency from any subcontractors because the schedule is driven by the performance of the trade foremen.

This policy does not imply that we push people around (that always backfires). It does demonstrate our commitment to a high level of project control and leadership by our Superintendents. It is imperative that this policy be carried out professionally. Construction schedules demand strong Superintendents who make and keep their commitments, and demand that their subcontractors and vendors do so also. This policy will benefit our subcontractors, gain the respect of their foreman and increase their profits.

Hoisting (For Internal Use)

Hoisting and handling subcontractors' materials will be carefully evaluated before the award of subcontracts. Overall efficiency suggests that each contractor be responsible to hoist and handle their own materials, and if we agree to do it, we intend to make a profit. Our efficiency is impacted when we hoist or handle for others, particularly when it is necessary to do it on overtime, since we need the equipment during regular hours. We are not usually compensated for the premium time, and we lose operator efficiency when they work long hours.

Subcontracting (For Internal Use Only)

Wherever possible, subcontractors will be awarded the full section of their specification because experience suggests that work in specification sections is more efficient when performed by a single contractor. If work is taken back from a subcontractor during buy-out, the purchaser must be certain it is appropriate and will create an overall savings. The Superintendent will be consulted about the efficiency and cost of the subcontractor's work that they will be required to perform before it is taken back from the subcontractor. If work is taken back from a subcontractor, it will be clearly itemized and copied to the Project Manager and Superintendent to avoid miscommunication, confusion, and time lost.

Figure 11.2 Sample subcontractor management policy for internal use.

How Much Equipment to Own

The first step in controlling equipment costs is to control the amount of equipment owned. The decision to purchase new equipment is basically made for two reasons: (1) to replace aging or less efficient equipment, or (2) for expansion. Both reasons are valid but must be considered judiciously because the company is committing money with limited assurance of future work. Work under contract, plus a reasonably certain backlog, will keep equipment busy for a year or more, but the equipment last multiple years. If a piece of equipment is not working at least 65% of the time that it is owned, it is generally cheaper to rent one as needed, as opposed to owning it.

Replacing Equipment

When considering replacing equipment, management must weigh whether the new equipment is really more productive than what they already have, and if so, by what margin? Other questions include the following:

a   Is the difference in productivity worth the investment?
b   If the existing piece of equipment is a maintenance headache, should the company consider investing in a complete reconditioning of the existing equipment and perhaps get three or four more years out of it?
c   Are the next two to five years of work a certainty? Can we keep the new piece of equipment busy?
d   Is the marketplace growing or shrinking at this time, and is it likely to change soon in either direction?

It is difficult to answer these questions with absolute certainty; however, these are questions that management needs to consider and weigh into the equipment purchase decision. The decision to purchase equipment is not easy. New equipment must produce profits over a number of years and not become a financial liability to the organization. The decision to buy means taking on additional costs and creating a necessity to get at least enough work to keep the equipment busy. Too often, contractors "want" to buy newer and bigger equipment, rather than having a real "need" to buy it. When they need a replacement, some contractors assume that bigger is better.

Purchasing for Expansion

The decision to buy additional equipment for expansion is usually made for one of two reasons: (1) new work is already contracted and there is no owned equipment available to perform the work; or (2) the contractor is in an expanding marketplace and wants to have the equipment on hand to perform the anticipated greater volume. The problem with buying or committing to more equipment in advance of getting the work is that the company is required to get more work just to keep the equipment busy and to justify the investment. In this situation, one can argue that the equipment may be running the company as opposed to the company running the equipment.

If the marketplace is getting stronger and growing, it may be reasonable to assume that the company will get their share of the growth, and therefore greater volume. However, the company's competitive position relative to their marketplace may not stay the same when the marketplace grows. Their regular competition may also have a bigger appetite and may be going after the work more aggressively than they did in the past. When a marketplace gets stronger it may also attract outside competition. When out-of-area contractors are drawn to a strong or growing marketplace they need to get a foothold and often bid very competitively to get their first jobs, and local contractors may react by bidding even more aggressively. Equipment becomes a burden when it forces a contractor to go after a greater amount of work at a time when they might have to bid work at a lower markup in order to get it.

If a company has contracted for more work than their current equipment can bear, the conditions of an expanding marketplace may still make purchasing risky. Once the existing work starts to finish up and the equipment used on those jobs becomes available, the company may have idle machinery. Increasing inventory of equipment should be very carefully thought out as it relates to the work, the marketplace, the firm's competitive position, and resources to perform the additional work while generating profit. Successful contractors have loaded up on equipment in good times only to be forced into difficulties by the equipment when the marketplace went back to normal. It is crucial to effectively calculate actual equipment costs and needs to maximize profits.

Calculating Equipment Costs

The subject of calculating and accounting for owned equipment cost is ignored by some organizations. Other organizations lose sight of this cost because they believe their accountants are adequately taking care of it. To estimate work, a contractor needs to know exactly how much equipment is going to cost, per unit time, including all maintenance and replacement costs. A contractor must have this information to accurately price the work at a profit. This concept sounds simple, but the calculations can be challenging.

One fundamental objective of operating a business is to produce net income. Net income results from selling a product or service to a customer and receiving more money than the total expense incurred to create the product or service. Equipment costs include an annual reduction in value or depreciation. Depreciation expense is the portion of the value of a piece of equipment (asset) that expires in each accounting period during the useful life of the equipment. This periodic cost requires no cash outlay, but is a continuous expense of operating the business because the equipment will need to be replaced at some point. Generally Accepted Accounting Principles (GAAP) rules provide a number of methods to calculate depreciation, none of which include the replacement cost of the equipment. Recovering replacement cost is a critical consideration for equipment-intensive contractors.

Time and Usage

The cost of owning equipment is a function of both time and usage. Some equipment may be busy all the time under normal one-shift-per-day conditions. Let's use the example of a rock crusher, which in this case works just about every weekday of the year. Assume a contractor were to purchase a new rock crusher for a multi-year project. The direct costs to the company during the first month of operation, assuming mobilization is charged separately, are fuel, insurance, and regular maintenance. These costs are easy to track monthly because they are incurred during each month of operation. As the project progresses and major equipment maintenance and spare parts are required, an allowance for the estimated costs of the repairs is included in the unit price of the equipment.

The allowance for maintenance and parts is an estimated cost that should be tracked and corrected occasionally to reflect what is actually occurring. However, the allowance needs to be applied as an actual cost each month, because when the crusher needs new belts and bearings, the cost for these parts is not incurred in the month the new parts are put in. The belts and bearings were consumed in prior months, so the future costs were incurred in those prior months and, in an attempt to reflect this reality, the estimated cost should be applied in advance of the maintenance event. The way to accomplish that is to charge an estimated amount monthly, since the exact cost is not known in advance. The estimated costs need to be applied in a timely manner to account for the true cost of the equipment, including future costs. Estimated maintenance costs should be updated periodically, at least annually.

Estimated regular and extraordinary maintenance costs are real and necessary to keep the equipment operable and in working condition; they must be included in the unit cost of owned equipment. If these costs are not charged to the unit costs, then the cost of some maintenance will come out of the company's profits. If the equipment will need major overhauls, such as engine replacements after two or three years, these costs have to be factored into the estimated maintenance and accounted for from the very first month the equipment is put into service. If that is not done, the company will be overstating their real profit by not charging the wear and tear of the equipment to the jobs it was used for, which creates a false economy.

Replacement Costs

In order for equipment-intensive contractors to be able to replace their machinery, they must include the replacement cost of the equipment in their unit prices. The replacement cost should not be confused with the original purchase price; it is in addition to the original purchase price. The replacement cost is calculated by determining the useful life of the equipment and estimating the future replacement cost (excluding salvage value) at the end of its useful life. The future replacement cost is divided by the useful life to get the monthly cost incurred. In this calculation, the time per year (in number of hours, weeks, or months) during which the equipment is anticipated to work should be used. If the cost is estimated assuming full-time work for an entire year, but the piece of equipment does not actually work full time, then the total replacement cost will not be fully recovered from the unit cost. The replacement cost represents the cost to the company of using and consuming the piece of equipment. If an organization uses the purchase price for replacement costs, as many do, they will not recover the additional funds (including inflation) required to replace the equipment in the future.

A contractor may ask, *"Do you really want me to charge my clients for next year's inflation when I'm only working for them this year?"* A lighthearted answer would be, *"No. You can always pay for it yourself."* If an equipment-intensive contractor does not charge replacement cost, they are paying for the privilege of being in business. They are consuming equipment at rates that are intended to repay the company for the amount they paid for the machinery. However, when the equipment will need to be replaced in a few years, the company will need more money than what was originally paid. To be self-sustaining, an operating business must replenish and regenerate itself from its operations, which includes the cost of ownership. It is an economic reality that inflation is a cost of doing business. For an equipment-intensive contractor, this means charging replacement cost to the equipment unit cost.

Economic analyses supporting a decision on equipment replacement are aimed at determining the equipment replacement interval that will yield the maximum profit on the investment (of the equipment). The period of equipment ownership that yields the maximum profit on the equipment investment may be considerably shorter than the useful life of the equipment. Equipment ownership costs, as the term implies, represent the total cost of owning the equipment. Although these costs are usually prorated on an hourly, daily, weekly, or monthly basis for estimating and accounting purposes, they represent considerable costs that will be incurred whether the equipment is actually working or not.

Equipment Costs Charged to Projects

The purpose of charging all equipment costs to a contractor's jobs and applying these charges on a monthly basis is to give the contractor a realistic picture of whether their company is making profit or losing money, in time to actually do something about it. Equipment is not an overhead cost any more than moving equipment back to the yard stops ongoing costs. Equipment is an operations or field cost.

To further expand on the concept of equipment costs and charges, consider a sample equipment-intensive contractor during a slow market period. With fewer jobs, the company has no place to charge ongoing costs for some of the time. During a slow period, there will be no fuel costs, and some maintenance can be suspended; however, insurance costs continue as do replacement costs, and when not charged to a project will have to be absorbed by the company.

Replacement is generally considered a function of usage and age; however, obsolescence can also cause replacement, so the timing of replacement is difficult to predict. If a contractor believes that downtime will extend the useful life of the equipment, then they can adjust the replacement cost as long as they factor in an amount for deterioration from storage and non-usage. Deterioration can be a costly factor because most construction equipment ages better in use than out of use. There can be significant costs associated with idle equipment.

Idle Equipment

When the market declines, taking on work in other geographic locations just to keep the equipment working is generally not a good business practice because the company often takes on too much risk. Additionally, taking on highly competitive work just to break-even is rarely justified, except for survival. One alternative is to liquidate some equipment; however, this decision must be considered in the context of the overall business, including new work anticipated. There is seldom a profit to be made in liquidating used construction equipment, although liquidation can reduce losses caused by the ongoing cost of idle equipment. Leasing out idle equipment is a possible alternative, but that may be difficult if there is a general slowdown in the market.

If nothing can be done to mitigate the loss from idle equipment, consider leaving it on the jobsite where it was used. One potential benefit is to avoid transporting expenses in the short term. The real costs of owning the equipment need to be charged somewhere, whether that is overhead or charged to the job. If left on the job, it serves as a daily reminder to the project team that the equipment is idle and may encourage team members to talk about where it should go next. Idle equipment on the job may discourage the project team from asking for more equipment than they really need in the future. The main benefit from this practice, however, is that if idle equipment makes one or more jobs show losses by the month, it simply points out that real costs are being incurred. If idle equipment is not charged to projects, it is possible for all jobs to be showing a profit on paper, while the real financial condition of the company is not as good as the paperwork is showing. This limited usage can be mistaken for profit.

Cash Flow

The majority of the cost of equipment ownership does not occur concurrently with the equipment's usage, so equipment intensive-contractors can have positive cash flow from projects that can be mistaken for company-wide profit. If the estimated replacement charges and extraordinary maintenance charges are spent or not placed in reserves, then funds will not be there when needed. During slow periods with a lot of idle machinery, an equipment-intensive contractor can incur losses but still have cash flow, which helps weather the storm, but is not profit. If the company uses the funds reserved for equipment replacement, there is not going to be enough money to replace the equipment when the time comes.

Since most construction equipment is purchased on credit, rather than cash, the equipment is expected to be paid off from future work. For equipment purchased on credit, interest costs are added to the formula as an expense. Since both interest and principal must be paid concurrently, cash flow during slow periods is affected. During idle times, if losses occur there may be negative cash flow. Depending on the length of a slowdown, there may not be enough cash flow for contractors to make the required payments of their equipment.

Equipment Obsolescence

Equipment-intensive contractors have another exposure that is not as apparent and often not planned for: equipment obsolescence. Companies incur a great deal of cost in replacing equipment as it wears out because broken-down equipment delay jobs, hinder progress, and can cost a lot of money. The productivity of equipment dramatically affects the profitability of the equipment-intensive contractors and is part of their competitive edge. Equipment productivity is critical to bidding and getting the work. As newer and more productive equipment enter the market and a contractor's competitors purchase it, the contractor can be forced into equipment replacement earlier than planned just to remain competitive. Obsolescence can occur well before machinery reaches its useful life. This is a difficult issue to overcome since it is very difficult to predict and plan for. Equipment obsolescence is therefore a risk of doing business for equipment-intensive contractors. The following example is intended to help define the concept of equipment obsolescence.

Mini Case Study #1

A well-established specialty contractor (HVAC) with aging duct-making machinery was facing new and unexpected competition from a start-up contractor who had modern duct-making equipment with the latest technology. The productivity of the new equipment allowed the start-up contractor to bid lower on every job of any size that came out during their first year in business. The stat-up contractor could undercut the established contractor's costs every time. The established contractor eventually decided the only way to remain in business was to replace their equipment with the more productive machinery their competition was using. Like most contractors, they had not reserved money for equipment replacement. They hadn't even accounted for it. The company's current financial statements reflected the bad year because of the new competition, so they could not secure the financing and were unable to purchase the new machines.

The established contractor was aware of obsolescence and the benefits of new technologies and kept abreast of the latest developments in their field, including computer-operated duct fabricating machinery. They knew that eventually some or all of their equipment would need to be replaced, but felt that most of the machines had a lot of good years left in them. Despite all of this previous knowledge, replacement costs were not included anywhere in the contractor's cost accounting or pricing of their work. By ignoring the real cost of replacing their equipment, the company was simply enjoying a "false profit." (Figure 11.3) Had the contractor accounted for realistic replacement reserves, they would have seen that their real profits weren't what they thought they were. Had they addressed obsolescence realistically, they may have planned for continual upgrading of equipment or at least measured how far the company was falling behind in technological developments. These considerations would have helped the contractor to quantify the risk of equipment obsolescence to the business. There is nothing inherently wrong with not replacing equipment at any given time. What is wrong in this case is for an equipment-intensive contractor to not account for, and include in their pricing, the total cost of equipment ownership which includes replacement cost.

Successful contractors understand that competition is a strong force in the construction industry. New businesses with better ideas are a reality in this industry; existing competition may gear up and tool up at any time to increase their market share; and out-of-town contractors are always on the lookout for new areas to expand into.

An equipment-intensive construction enterprise that does not concern themselves with the effect of obsolescence on their business can place themselves at great risk. Reserving the anticipated cost of "keeping up" may not be enough if a company is not be able to gear up fast enough when their competitive balance shifts rapidly. It is advisable for contractors to use their equipment replacement reserves to keep up with national standards, not just local standards, because the industry is mobile and unanticipated competition can originate from anywhere.

Figure 11.3 By ignoring the real cost of replacing their equipment, the company was simply enjoying a "false profit."

Daily Incurred Replacement Cost

The entire future replacement cost of equipment, including the costs due to inflation, wear and tear, and obsolescence, will all need to be paid at some point whether that is accounted for or not. Replacement cost is a real cost of doing business. It is a cost that is incurred each day. The following is a hypothetical example of the concept of daily incurred replacement cost.

Mini Case Study #2

An equipment operator decides to go into the dirt-moving business and buys a $100,000 bulldozer (ignore the exact amount; this number is used simply for easy math to illustrate the concept). Now the operator is faced with the question of how to account for the ownership of this piece of equipment in years to come. For this example, assume that the contractor bought the equipment for cash from personal savings and that the bulldozer will last for five years, at which time it will be worn out and have zero salvage value. There are a number of ways to account for depreciation; in this example we use a straight-line depreciation over five years, which results in depreciation of $20,000 per year.

The new dirt-moving business recovers usage costs during the next five years, including $20,000 depreciation in their accounting for equipment ownership. Where will they be in five years? The short answer is that they will no longer have their $100,000 because they spent it to buy the bulldozer in the first place. They will not have the bulldozer either, since it is worn out and has no salvage value. And they may be out of a job without the piece of equipment.

What happened to the $100,000? The answer is that it has been consumed by the business. The contractor expended revenue on salary and operating costs and made reasonable profits during those years while depreciation allowed $20,000 of income a year without corporate tax. The contractor having earned the profits (both real and "false profit") took most of it out of the company and expended it on lifestyle costs. The replacement bulldozer today costs more than the $100,000 it cost five years ago. It now costs $150,000. To stay in business the contractor had to borrow the $150,000. In this situation, the contractor is not only short of the $100,000 that they started with five years ago, but they are also in debt of another $150,000. The main point here is not the *necessity* of equipment replacement, it is the *real cost* of equipment replacement.

While the example is simplistic in nature, it provides food for thought. Replacement costs to be incurred five or more years in the future would be calculated at the present value of the future cost, and tax considerations would impact the calculation. Nevertheless, replacement costs will become due at some point and will be more than the original purchase price, so a contractor cannot rely on allowable depreciation alone to accurately account for the cost of equipment ownership. The company will go deeper into debt by at least the rate of inflation every time it has to replace a piece of equipment.

The contractor in the case study never considered setting aside the $20,000 a year depreciation, thinking that the money used for the purchase was his to begin with. If he had set aside the $20,000 yearly, he would have at least had the $100,000 available to purchase the replacement bulldozer. But that is not even enough because the new bulldozer costs $150,000. What happened here is that he "consumed" the bulldozer during the operation of the business so the extra $50,000 was, in effect, "given" to the customers. The $100,000 depreciation was included in the hourly costs and collected, but the contractor did not put that money aside for the replacement he knew would be needed. These funds were collected and considered available, which is why we refer to it as "false profit." If he did retain the $100,000 he would still have been short the $50,000 inflation which we call "false profit" because it is essential to replace the means by which the continuation of profit is earned. The message is that the replacement cost of equipment is a cost of doing business. Irrespective of what it is called, it must be retained for replacement.

Summary on Equipment Management

When a company does not account for the real replacement cost of equipment, profits are exaggerated, which gives a false picture of where the organization is and of where they are headed. While funds reserved for equipment replacement costs are not tax deductible, they are clearly a cost of doing business. There is a lot of debate on this subject from accountants and tax experts, and it would be beneficial for the industry to unite and determine what the optimal method is for all parties considered.

Economic conditions in the construction industry can make it hard to distinguish if a piece of equipment is a liability or an asset. Ignoring the real replacement cost of equipment necessary to remain in business can cause an organization to operate in a false economy, go further into debt over time, and for many contractors create serious long-term financial problems. Costs that will be incurred in the future and will become due in subsequent accounting periods are no less real than costs that are incurred and due in the current accounting period. It is just harder to recognize and account for them. Equipment cost management is a critical concept in the construction business, particularly for equipment-intensive contractors.

Chapter Summary

Subcontractors and equipment represent a large portion of field costs, and the quality of their management is critical to a construction entity's success. The information shown in this chapter describes tried and proven successful methods, procedures, and policies. The reader is encouraged to modify these based on their individual needs. Consistency throughout an organization is not only professional; it is the best way to accurately monitor and measure performance. The successful contractor takes the time and effort to determine what works best for them, and then establishes the way their company will consistently, profitably, and safely perform. Next, they regularly measure the results to ensure continuous improvement.

Review Questions. Check All That Apply.

1   The greatest potential for project is when each subcontractor's work is:
a   Around their average size
b   Typical of their larger projects
c   Typical of the smaller projects
d   None of the above

2   What do the authors advise should be done with a well thought out written subcontractor management policy?
a   Review it annually
b   Institute it company-wide
c   Refer to in the marketing brochure
d   All of the above

3   What percent of the time do the authors say a piece of equipment must work to be owned rather than rented?
a   50% of the time
b   60% of the time
c   70% of the time
d   None of the above

4   Equipment replacement is generally considered a function of age and usage; however, what else do the authors say can cause replacement?
a   Available funds
b   Desire to upgrade the fleet
c   The need for a tax deduction
d   None of the above

5   When a company does not account for the real replacement cost of equipment:
a   It is accounted for in overhead
b   The company is less competitive
c   Profits are diminished
d   None of the above

Critical Thinking and Discussion Questions

1   Discuss the purpose for a subcontract management policy.
2   Explain why you would institute a subcontractor management policy.
3   Explain the author's position on how much equipment to own.
4   Discuss the concept of the replacement cost of a piece of equipment.
5   Discuss the significance of accurately understanding the replace cost of equipment.

12 Customer Service

Customer service is a business's ability to satisfy its customers (Figure 12.1). Great customer service is when a company exceeds the customer's expectations. Given the construction industry's increasingly competitive market, some argue that we are in the midst of a business revolution revolving around customer service. Organizations are attempting to defend their market share, and increase it by increasing customer satisfaction through a focus on the quality and service provided. This movement has produced significant benefits.

Great customer service begins with considering everyone you work with to be a customer, including subcontractors, general contractors, project owners, designers, government, the general public, and many others. Each of these stakeholders has different needs that must be understood, appreciated, and satisfied.

There are many challenges to achieving great customer service. The first challenge is the intangible nature of the process. Although a construction product

Figure 12.1 Customer service is a business's ability to satisfy its customers.

DOI: 10.1201/9781003229599-15

(e.g., building, bridge, plant) is tangible, the process of getting it designed and built on time, installed correctly, and then running smoothly involves a lot more than the product itself. The intangible aspects of the project delivery process can make or break a company's success. Prospective customers cannot evaluate the intangible nature of the delivery process in advance. There is no working product until it is delivered, and the customer won't know how well it performs until it is actually put in use. Contractors can have all the tangible elements of customer service in place, from advanced technology to great engineering to efficient installation, but if customers are not satisfied with the way their transaction and interaction is handled, they may not be back.

According to Theodore Levitt, Harvard Business Review, No. 81306:

> The most important thing to know about intangible products is that the customers usually don't know what they're getting until they don't get it. Only then do they become aware of what they bargained for; only on dissatisfaction do they dwell on the process and related interactions. Satisfaction is as it should be, mute. Its existence is affirmed only by its absence.

The involvement with a customer for the length of time it takes to complete a project (several months and often years) provides an opportunity to do 99 things right and one thing wrong, and be noticed and remembered for the one thing done wrong.

A second challenge to achieve great customer service is the commodity mindset. The "commodity mindset" of some customers is that contractors are merely performing a function in building what they need. Doing a good job is taken for granted by most buyers of construction services. They assume that all contractors would provide a similar product. However, the client has to live with the project team for the life of the project, experience the changes in design which are impacted by their selection of architects or engineers, assess the product as it develops – compared with their expectations, and pay invoices they often don't fully understand.

Third, the fact that projects have multiple touch points is another challenge to achieving great customer service. The entire organization must "buy-in" to the importance of customer service. We can do a great job technically, but if our customer has a less-than-positive experience in dealing with any of our team members, the customer may very well believe we did not do a good job. Every customer point-of-contact throughout the timeline of the project must be positive and cooperative (or at the very least, neutral) or our customers may not be satisfied. Research suggests we are in an "experience economy" in which customers expect good products and services from everyone but buy from purveyors that make the "purchase experience" (from qualifying for the project all the way to closeout) a positive – even exciting – experience. Keep in mind that one of the easiest ways to turn a service into an experience is to provide poor service, thereby creating a memorable encounter of the unpleasant kind.

A Case Study on Customer Service

To help understand the multiple aspects of managing customer relations and creating a strong customer service perspective within your organization, let us follow a specialty contractor through his discovery of the issues and the resulting customer service changes he made within his organization. Depending on the size and individual needs of your company, you may want to adopt only some of the methods and processes from the following case study and certainly feel free to alter and modify any of them to fit your particular needs. Also keep in mind that influencing anyone's attitude is not easy and takes time. However, the customer service endeavor is well worth the time because improving customer relations often improves profitability and repeat business.

The owner of a specialty contracting firm was great at selling his company's services to all the general contractors in his region. He contacted them frequently and played golf with some of them regularly. However, he was not getting the repeat business he expected. To investigate the reason for the lack of repeat business, he set up a series of lunch meetings with key clients. At first, clients wouldn't say much, but after considerable cajoling, he found out that one of his most profitable project managers was hard to work with, as were some of his other employees. Some were reported to be rude to the general contractors' employees or abrupt with the owner's representatives, and others were said to be unprofessional with their subcontractors and vendors. Additionally, some of his people were said to have disregarded the needs of project neighbors or the general public.

The contractor further explored this last complaint about project neighbors and the public to discover that on various projects his employees had created noise, dust, or disturbance that the customer considered excessive or unnecessary, in areas where project specifications specifically prohibited that. There were complaints of his employees parking outside the areas allowed for construction use and, in some cases, even in the customer's reserved parking spaces or on the grass or landscaped areas. The complaints ranged from trash blowing over the customer's non-construction occupied property to employees using loud, foul language in customer-occupied spaces. The list was long and embarrassing.

It became apparent that the company's foremen and tradesmen had little guidance and were uncooperative, turning customers off and leaving the customers (in this case, general contractors) with a negative feeling about working with their company. Some customer representatives and project owners reported that they sometimes minimized job site visits because they felt they were not welcome on their own project.

Our case study contractor knew he had a serious problem but wasn't sure where to begin. He asked himself the following questions:

- How do you change the ingrained culture of the company?
- What is great customer service in our business?
- Exactly who really are our customers?
- What do our customers want?

- What are we delivering now?
- Where are we falling short?
- Is improved customer service going to make us money or cost us money?

He concluded that in order to make a start in changing the culture of his company he would need to become a customer-satisfying organization, stop putting all his emphasis on efficiency and the bottom line, and start adding some emphasis on constantly improving customer service. He came to understand that his marketing of his company's services should not focus on what his company needs, but instead on what his customers want. A genuine commitment to customer satisfaction begins at the top, starting with him.

The Customer Service Group

Our contractor knew that his team would need to be part of the change process. He gathered his key employees to discuss the problem and to help design the solution. The group agreed that they needed to create a customer service strategy and plan. The contractor felt that all aspects of the company would need to be represented, so the newly formed committee was composed of the VP finance, VP marketing, the project manager who was reported as hard to work with, an internal engineer, and a field foreman. *(Readers should consider their own configuration within their unique organization.)*

The group first asked the following question: Who is our customer and what do they want? General contractors are the most frequent customer of the specialty contractor. The project owners are another key customer because they live with our work for many years. There are many additional customers to be addressed too. The group decided they needed to define customer service for their business and establish a customer service philosophy and values. They set three simple performance goals.

1   Consistently give the highest level of service that the company is capable of providing.
2   Be recognized as the best customer service company in their market.
3   Be identified as a leader in customer service in their industry.

The committee set a goal of a 10% annual improvement in customer satisfaction as measured by the number and severity of customer complaints recorded through a new feedback system that they would develop. *(Readers may want to consider a different percentage that they believe their organization can achieve.)*

When it comes to specific action items, the group decided they would conduct a cost-benefit analysis at the end of each plan year to determine if the program is accomplishing the three stated performance goals, conduct *toolbox talks* that address customer service issues, and design a reward process to celebrate individual and team customer service successes. In the first year, the group took the following actions:

- Prepared a customer service plan.
- Included customer service education and training in company training programs and retreats.
- Established pre-project and post-project customer interviews and learned from the results (examples will be presented in the next section of this chapter).
- Implemented celebrations of service success.
- Developed the following mission statement for the company:

> We build and install systems for people not corporations, and we treat our customers as we would like to be treated. Our customers are always welcome on the job because it is their project, and we appreciate the opportunity to show off the work we do on their behalf.

The group wanted to develop valid processes that gather information from customers and ask the right questions to the right people so they can receive effective feedback. They decided to visit two or three clients for a short discussion about what they expect of the PMs, foremen, and tradesmen during the construction process. This simple idea was very effective because they learned what actually matters to their customers. The clients were thrilled to participate and were left with a good feeling about the company. It was so successful that they conducted several additional customer visits.

Customer Questionnaires

Here are the pre-construction and post-construction questionnaires the team came up with (Figures 12.2 and 12.3). As you work with your existing customers you will add, remove, alter, or substitute other questions that are more relevant to your clients.

Customer Service in the Company Culture and Tied to Business Outcomes

The group decided that keeping customers satisfied during construction is a huge issue because it affects the company's ability to avoid disputes, sets the stage for argument-free change orders, and generates positive recommendations for future work. They decided they need to care about the customer's wants, learn to "say yes" whenever possible, and treat the customer as they would like to be treated. The group recommended the following steps to infuse the company culture with a "customer satisfaction" attitude.

- Add customer service abilities to compensation metrics.
- Review "customer experience" project scores regularly.
- Screen prospective employees for customer service values.

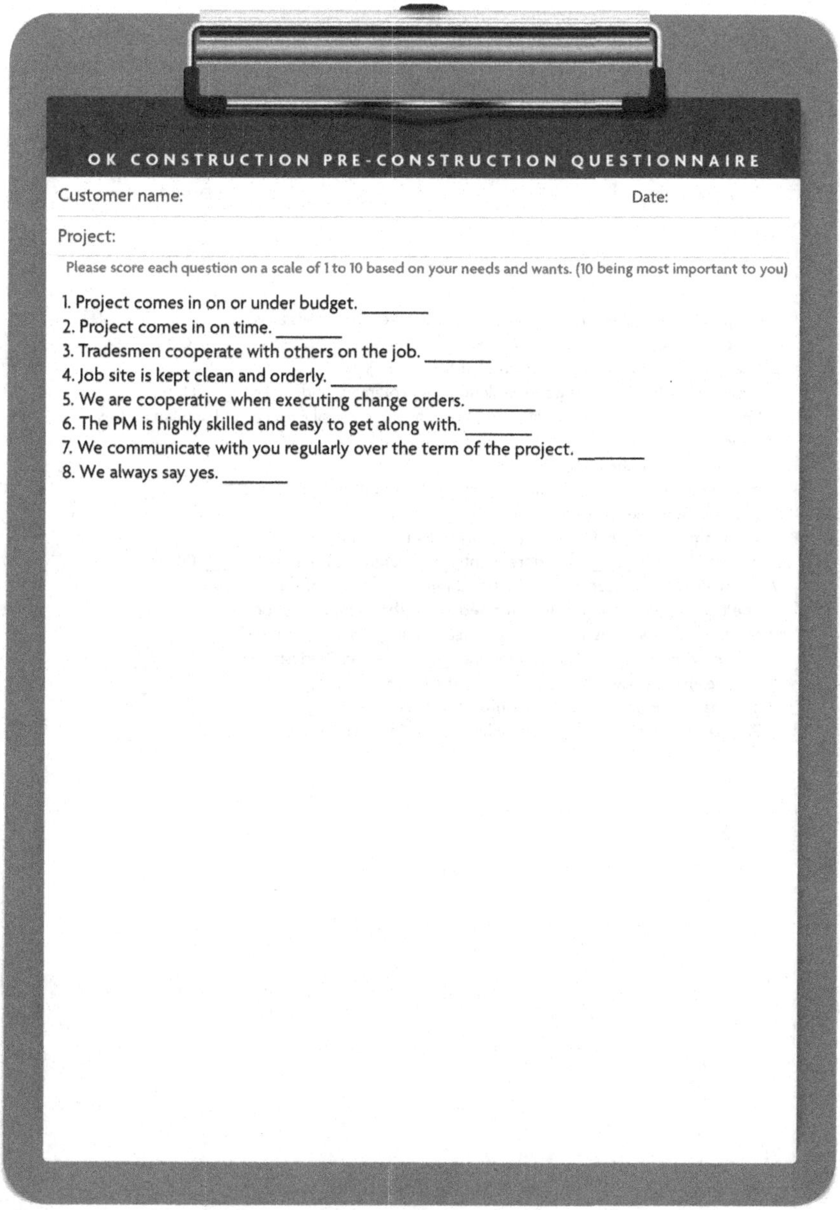

OK CONSTRUCTION PRE-CONSTRUCTION QUESTIONNAIRE

Customer name: _____   Date: _____

Project: _____

Please score each question on a scale of 1 to 10 based on your needs and wants. (10 being most important to you)

1. Project comes in on or under budget. _____
2. Project comes in on time. _____
3. Tradesmen cooperate with others on the job. _____
4. Job site is kept clean and orderly. _____
5. We are cooperative when executing change orders. _____
6. The PM is highly skilled and easy to get along with. _____
7. We communicate with you regularly over the term of the project. _____
8. We always say yes. _____

Figure 12.2 Sample pre-construction questionnaire.

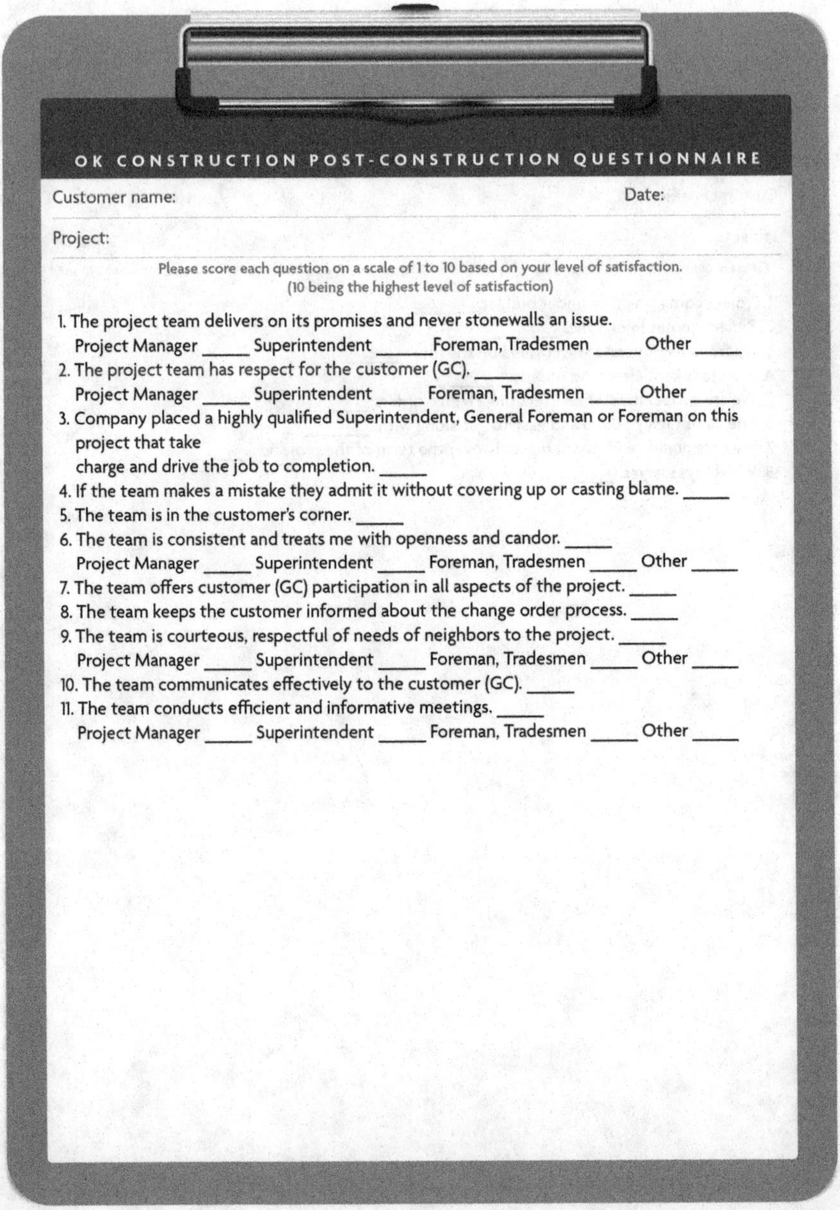

OK CONSTRUCTION POST-CONSTRUCTION QUESTIONNAIRE

Customer name: Date:

Project:

Please score each question on a scale of 1 to 10 based on your level of satisfaction.
(10 being the highest level of satisfaction)

1. The project team delivers on its promises and never stonewalls an issue.
   Project Manager _____ Superintendent _____ Foreman, Tradesmen _____ Other _____
2. The project team has respect for the customer (GC). _____
   Project Manager _____ Superintendent _____ Foreman, Tradesmen _____ Other _____
3. Company placed a highly qualified Superintendent, General Foreman or Foreman on this project that take
   charge and drive the job to completion. _____
4. If the team makes a mistake they admit it without covering up or casting blame. _____
5. The team is in the customer's corner. _____
6. The team is consistent and treats me with openness and candor. _____
   Project Manager _____ Superintendent _____ Foreman, Tradesmen _____ Other _____
7. The team offers customer (GC) participation in all aspects of the project. _____
8. The team keeps the customer informed about the change order process. _____
9. The team is courteous, respectful of needs of neighbors to the project. _____
   Project Manager _____ Superintendent _____ Foreman, Tradesmen _____ Other _____
10. The team communicates effectively to the customer (GC). _____
11. The team conducts efficient and informative meetings. _____
    Project Manager _____ Superintendent _____ Foreman, Tradesmen _____ Other _____

Figure 12.3 Sample post-construction questionnaire.

- Train new employees to understand who our customers are and what they expect.
- Ensure courtesy in answering phone calls and responding in a reasonable time.

To disseminate the initiative internally to all team members, the group listed ways in which great customer service will enhance business outcomes, including:

- Enhances reputation of the company, leading to referrals.
- Promotes repeat business.
- Increased efficiency by reducing on-the-job conflicts.
- Facilitates a cooperative change order procedure.
- Improves productivity through a positive attitude during the construction process.

Change Orders – A Customer Service Issue

What do architects and engineers hate, owners mistrust, and everyone blames on the contractor? You guessed it: change orders! Change orders are not only a field management concern and a financial concern, but also a major customer service issue. The best way to demonstrate that your company is not responsible for – nor causes – change orders, is to clearly explain the change order process including why change orders exist, what causes them, and that they are a standard element of the construction process. Properly introduced before the start of construction and managed adequately, change orders can be neutralized and potentially perceived as a service to the customer.

Managing change orders begins with understanding that when there are changes to the work during construction, they often disrupt production and schedule. Being compensated for the changes seldom covers the cost of disruption. Most designers and owners resist change orders and often blame contractors for initiating the change. However, change orders are often the result of changed conditions, designers changing their mind about some details, or the owner adding something to the project. Contractors are often accused of going after change orders that they are not entitled to, but that is actually the exception, not the rule. What does happen is a poorly managed project resulting in cost overruns that an unscrupulous contractor may attempt to be paid for, using change orders as the vehicle. This exception reflects poorly on the entire industry.

Enlightened contractors consider change orders a distraction because, as research indicates, performance and schedule are significantly improved on jobs that have fewer change orders. If there is a changed condition or the project owner adds scope, a contractor should expect to be paid for it. However, the cumulative production and schedule disruptions from multiple change orders can get out of hand and drive up project costs considerably. Seldom is this ripple impact cost included in the change orders because it is almost impossible to calculate in advance; it also cannot be anticipated during the early change orders. The cause of most change orders is outside the influence of the contractor, and most change orders cannot be prevented by the contractor. A poorly designed project, or one where the design is incomplete, inevitably results in multiple change orders. Unfortunately, this condition, like most change orders, is often blamed on the contractor instead of the designer, which hurts the relationship between the contractor and the project owner.

In our experience, projects with hundreds of change orders are much more likely to underperform or lose money, and in a number of cases have put the contractor out of business. As the number of change orders mounts, it becomes harder and harder to negotiate the price of the new work and get change orders approved. Multiple change orders are clearly not advantageous. Fewer is better. Therefore, the enlightened contractor prefers that change orders do not occur, and when they do, the cause should be clearly identified. If this is made clear to designers and owners in advance, it can help set realistic expectations at the onset of the project, and owners will be more prepared to understand the real cause of each change order as it develops.

A Change Order Policy

The contractor needs to be proactive with change orders, beginning before construction is underway (Figure 12.4). It starts with having a company-wide change order

Figure 12.4 The contractor needs to be proactive with change orders.

policy that is widely circulated to owners, designers, and field and office employees. A change order policy explains to the owner and designer the contractor's expectations and position concerning change orders, and how the process works most efficiently. It demonstrates that the contractor's preference is to not have change orders, which disarms potential future accusations that change orders are being caused by the contractor. A sample change order policy is provided in this chapter as a guide that can be customized to individual contractor needs (Figure 12.5).

The first step, and perhaps the most important step, for a successful change order policy is for company managers and supervisors to understand that the policy is not intended to cause change orders, but to manage the process, explaining that the company's position is that fewer change orders are preferable, and also the reasoning behind that. This is critical because it is opposite to what a large number of owners and designers believe, causing them to be defensive and sometimes untrusting when dealing with change orders. Explaining that to managers and supervisors, including the fact that they are expected to actively assist owners and designers in avoiding change orders, is the purpose of the detailed instructions provided before the short policy document itself. That sets the stage for the very legitimate position that the contractor also intends to be compensated for changes to the scope of the work that they based their price on. Let's call our example company *OK Construction*; their instructions for change order management are presented next, followed by their sample change order policy.

Instructions for Change Order Management *(for Internal Use Only by OK Construction Managers and Supervisors)*

There is a prevailing belief in the industry among owners and architects that contractors look for and cause change orders, as a secondary source of profit. OK Construction does not solicit change orders or rely upon them as a source of profit. The purpose of this change order policy is to clarify for the designers and owners we serve, and their representatives, our approach to avoiding contractor-initiated change orders. Moreover, OK will efficiently process owner-initiated or designer-initiated change orders should those be needed.

Changes to the original scope of work have very real ramifications on the efficiency and schedule of the project. Continuous progress and productivity are key to the successful completion of the project and rely on timely decisions and the active participation of the owner and designer. Construction is a sequential process. When work is performed out of sequence, areas are left uncompleted, or rework is caused by lack of information, and progress and productivity are dramatically impacted. This impacts efficiency and ultimately the bottom line. Even when compensated for the extra work, we often cannot regain the loss of efficiency and productivity. If we are denied the resulting extensions of time or compensation for delay, we are forced to absorb overtime costs to maintain schedule. Contrary to popular belief, projects with large numbers of change orders turn out with worse profit for the contractor, and worse schedule outcomes, compared to

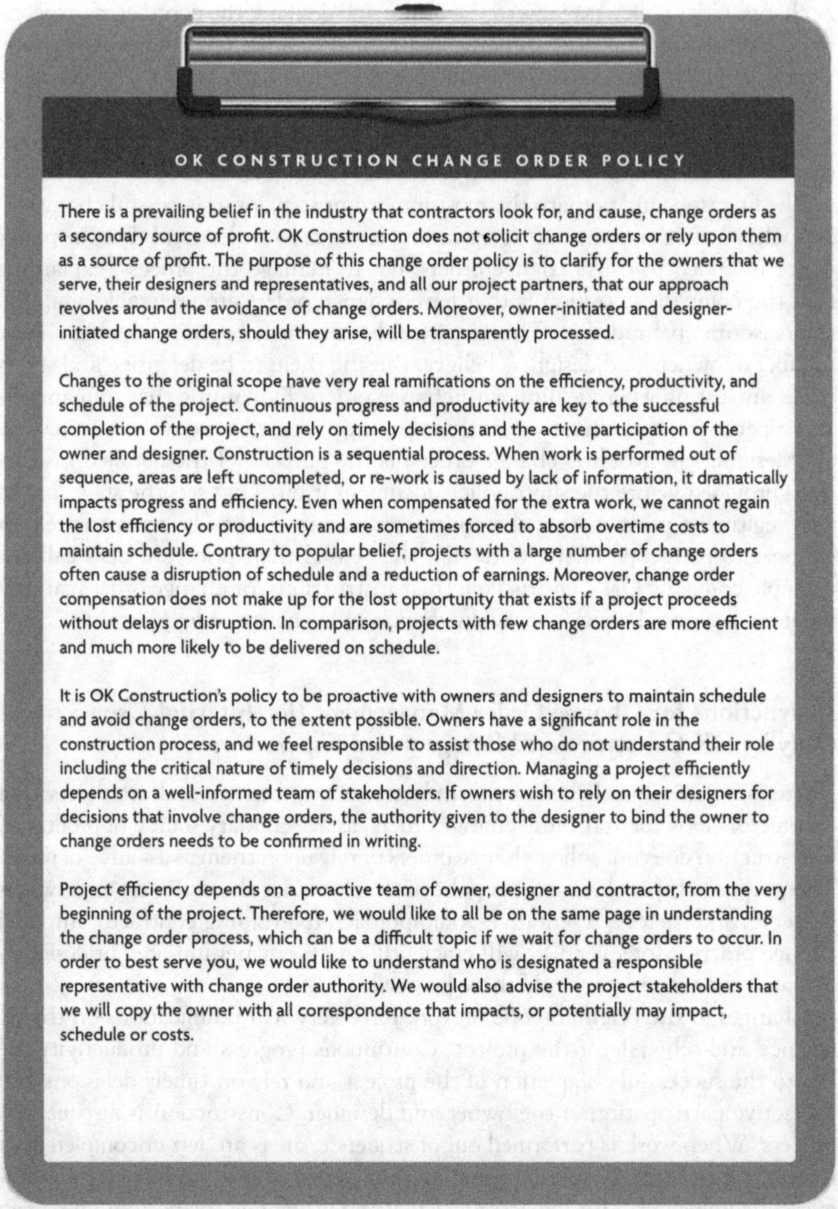

OK CONSTRUCTION CHANGE ORDER POLICY

There is a prevailing belief in the industry that contractors look for, and cause, change orders as a secondary source of profit. OK Construction does not solicit change orders or rely upon them as a source of profit. The purpose of this change order policy is to clarify for the owners that we serve, their designers and representatives, and all our project partners, that our approach revolves around the avoidance of change orders. Moreover, owner-initiated and designer-initiated change orders, should they arise, will be transparently processed.

Changes to the original scope have very real ramifications on the efficiency, productivity, and schedule of the project. Continuous progress and productivity are key to the successful completion of the project, and rely on timely decisions and the active participation of the owner and designer. Construction is a sequential process. When work is performed out of sequence, areas are left uncompleted, or re-work is caused by lack of information, it dramatically impacts progress and efficiency. Even when compensated for the extra work, we cannot regain the lost efficiency or productivity and are sometimes forced to absorb overtime costs to maintain schedule. Contrary to popular belief, projects with a large number of change orders often cause a disruption of schedule and a reduction of earnings. Moreover, change order compensation does not make up for the lost opportunity that exists if a project proceeds without delays or disruption. In comparison, projects with few change orders are more efficient and much more likely to be delivered on schedule.

It is OK Construction's policy to be proactive with owners and designers to maintain schedule and avoid change orders, to the extent possible. Owners have a significant role in the construction process, and we feel responsible to assist those who do not understand their role including the critical nature of timely decisions and direction. Managing a project efficiently depends on a well-informed team of stakeholders. If owners wish to rely on their designers for decisions that involve change orders, the authority given to the designer to bind the owner to change orders needs to be confirmed in writing.

Project efficiency depends on a proactive team of owner, designer and contractor, from the very beginning of the project. Therefore, we would like to all be on the same page in understanding the change order process, which can be a difficult topic if we wait for change orders to occur. In order to best serve you, we would like to understand who is designated a responsible representative with change order authority. We would also advise the project stakeholders that we will copy the owner with all correspondence that impacts, or potentially may impact, schedule or costs.

Figure 12.5 Sample change order policy.

projects with fewer change orders which are more efficient and able to be delivered on schedule. In addition, change order compensation for multiple changes does not make up for the lost opportunity that exists if a project proceeds without delays or disruption.

At the first project meeting (usually a pre-construction meeting), the OK Construction's project manager or superintendent will provide the owner and designer a copy of our change order policy. They will discuss the topic of handling change orders on the project with the owner directly, clearly, and professionally by utilizing the following procedure:

1   The discussion will be directly with the owner, not through the architect or engineer, but in the presence of the architect or engineer. It must be during the formal meeting while everyone is attentive – not a side conversation. The objective is to get the questions and answers into the meeting minutes.

2   Ask: "Who can authorize change-order work?" which usually confirms that only the owner has final and ultimate approval. It should also be clarified that the designers do not issue final approvals, because if they did, it would not be necessary for the owner to sign change orders. Unless the owner states for the record that the architect's approval of change orders is final and binding on the owner, ask the following question, "During the course of construction, should we proceed with change order work without owner authorization?" The answer is typically "No." At that point, address the architect or engineer by stating, "We will, therefore, wait for change orders signed by the owner before proceeding with any extra work no matter who orders it." You need to establish that we cannot accept direction to perform extra work without an assurance that we will be compensated for it. State that

We are sometimes directed to perform work outside the scope (or that changes the scope) with the designer saying it is not a change in scope. Then we are supposed to debate later with the same designers to get paid for that work.

We must attempt to demonstrate to the owner the unreasonableness of the position we are put in if we are to present the change of scope to the same person who already had declared that it was not outside of the scope when they directed that work to be performed. It is imperative that we demonstrate from the onset the absolute necessity of owner participation in disputed work. Otherwise, we state we will be unable to perform any extra work even if it delays the project. The objective is to establish an expeditious, binding, owner-involved change order process in advance of construction, because it is in the best interest of the project and all parties involved.

3   The change order process often does not include actual signed change orders because the designers often say that the paperwork takes a long time. That can be problematic. So, let's attempt to establish handwritten change orders

and/or signed field orders by the owner's authorized representative as binding. No matter what the results are, we establish that we intend to present changes of scope to the owner and expect a timely reply. If they do not want us to proceed with the work, they should say so. Owners often say they are unable to do this because they may not understand the nature of the work in question. This continues the discussion in the intended direction because it demonstrates that the owner must rely on their designer in technical areas. That is reasonable and understandable. However, we do not intend to be directed by the designer if they have no authority to approve the additional scope of work. Our position is: if the owner does not understand a technical issue on a field order that we send to them, they can say so, and we will not proceed with the work until it is clarified to their satisfaction. This usually leads to a discussion around the necessity of a rational change order management policy for the project.

4   Timing is crucial, so attempt to establish that all open change orders must be cleared at each monthly project meeting which the owner attends. Owners may say their representative at the meeting does not have the authority to approve change orders. The entire communication is an attempt to clearly demonstrate that we do not want to be directed to perform extra work, while no one with authority to approve compensation is available or even aware of the circumstance. Explain that active owner participation leads to avoiding disputes that are outside of our control. It is our position that the owner must be available for change order approval, on a monthly basis at a minimum, or they must authorize someone to bind them. Often an amount is established where the designer can bind up to a limit, the owner's representative up to a limit, and only the owner after that. This is acceptable.

5   It is helpful to the designer for you to state that "If the designer gets some authority to pay for extras, they are better equipped to manage the project in a fair and impartial manner, which is the basis of our contract with the owner."

6   The meeting minutes, unless prepared by us, seldom include the details necessary to clarify our position on change orders and extra work. As soon as the minutes are published, complete clarification of our position is to be sent to whoever wrote them, and ensuring the owner is copied on this communication. This also demonstrates our intention to copy the owner on all money and schedule issues, no matter what.

7   Even when the discussion is not entirely successful in establishing an improved change order management process, it communicates, and hopefully clarifies, the process for the owner and helps establish that we are not the primary cause of extra work disputes.

8   A change order meeting schedule should be agreed upon during the first project meeting.

Conclusion: Make Great Customer Service Your Primary Product

We can learn from a customer service survey conducted by Dimensional Research, an independent market research company, that surveyed over 1,000 individuals who had extensive experiences with the customer service activities of midsized companies. The survey quantified the long-term impact of customer service on business results.

Customer service impacts revenue. Participants ranked customer service as the #1 factor impacting vendor trust. Sixty-two percent of business-to-business and 42% of business (direct)-to-customer purchased more after a good customer service experience. Sixty-six percent of business-to-business and 52% of business-to-customers stopped buying after a bad customer service interaction.

Customer service experiences have a long-lasting impact. Twenty-two percent continue to seek out vendors two or more years after a good experience. Thirty-nine percent continue to avoid vendors two or more years after a bad experience. Customer service experiences are spread widely – especially the bad ones. Ninety-five percent share bad experiences and 87% share good experiences with others. Business-to-business most likely share their customer service stories.

What specifically made customer service interactions good?

"The problem was resolved quickly."

"The person who helped me was nice."

"The problem was resolved in one interaction – no passing around to multiple people."

"The outcome was what I was originally hoping for."

Among participants that had experienced good customer service, most of them (83%) reported that their behavior changed in some way as a result. The most common way that their behavior changed was that they purchased more from that company (52%).

Seeing your company as an organization that satisfies its customers can pay off significantly. A portion of our construction industry still has a "commodity mindset." So, in this market environment you need to know how to become the contractor of choice. How to stand out among all the rest. Many contractors are competent and can do the work, making it very difficult to display distinctive competence to distinguish your firm as considerably different from competitors. Why is one contractor invited to bid and not another? The contractor who creates a cooperative, positive, professionally respectful experience for the general contractor (GC), owner, other trades, designers, and project neighbors throughout the term of a construction project will often be on the preferred list of bidders.

Most contractors spend time thinking about efficiency and quality, which they should. However, to become the "contractor of choice" they have to recognize the importance of the customer's experience during the term of a project. Some contractors make the error of considering the owner or GC adversaries and potential obstacles to the contractor's profitability. These are ill-advised "beliefs" that some wrongfully consider the normal environment of the construction industry.

It is all about cooperation. In this "evolving" business environment, the future of your company depends on your ability to cooperate and work as a team. This will continue to be the case with alternative project delivery methods such as construction management at risk (CMAR) and design-build (DB) increasing in market share compared to the traditional design-bid-build project delivery. Contractors are being selected not just based on price, but also based on their qualifications including the owner's past experiences with the contractor. We are no longer just in the construction business. To be continuously successful, you need to be in the customer creation and customer satisfaction business as well. The program outlined in this chapter is just the beginning.

It takes time to effect a change like this in an entire company's view of itself and its mission. The key to success is in the strength of your commitment. If you and your senior management truly believe that *servicing customer's needs* must be part of your new business, this program will imbed that belief in the entire company, increasing cooperation and profit, perhaps even consider reflecting this belief in your mission statement, if it's not there yet.

Review Questions. Check All That Apply.

1   The authors say that the first challenge in achieving great customer service is:
    a   A huge issue
    b   Finding the right consultant
    c   The intangible nature of the process
    d   All of the above

2   A Harvard Business Review article says customers usually don't know what they're getting until:
    a   The project is completed
    b   The project is underway
    c   They don't get it
    d   They see it

3   When do the authors say is the time to demonstrate that your company is not responsible for nor causes change orders?
    a   As soon as the first change order arises
    b   Before the start of construction
    c   At contract signing
    d   None of the above

4   Continuous progress and productivity rely on:
    a   Timely decisions

b   Active participation of the owner
c   Active participation of the designer
d   All of the above

5   What percentage of bad customer service experiences do the authors report are spread widely?
a   65%
b   75%
c   85%
d   95%

Critical Thinking and Discussion Questions

1   Explain the purpose of a customer service group or committee.
2   Discuss the process used by the contractor in the case study on customer service.
3   Why are change orders a customer service issue?
4   Discuss the advantages of a change order policy.
5   Discuss the long-lasting impact of the customer service experience.

Part IV

Setting Up the Business for the Future

13 Construction Market Cycles

Not everyone in the construction industry fully understands the cyclical nature of the construction market. The construction market in the US is cyclical, and has been for a while, since at least WWII. The construction market follows the US economy almost in lockstep, except that it lags behind it. When the US economy cycles down, the construction market continues to grow or flatten out for 12–18 months, and then cycles down. The reason for this lag is that the majority of the construction projects in progress and those already funded usually continue through completion. The average length of construction projects, averaging at about a year or so, provides continued revenue for some time. During this lag, planned projects begin to be defunded and fewer new projects are planned, so the backlog to replace the work that is being completed eventually falls off.

When a US economic downturn bottoms out and begins to recover, the construction market lags the economic recovery and continues to decline. The reason for this lag is similar to the discussion in the earlier paragraph: the economic recovery stimulates new construction, but it takes time to decide which new projects will move forward, and additional time to design those projects. It also takes time to fund them, to seek approvals and proposals, and to contract for the projects to be constructed. As the construction market begins to grow again, it usually takes from 12 to 18 months for it to rebound to its pre-downturn size. Note that both the lag at market decline and the lag at market recovery are average values and vary widely depending on the length of the down market and the amount (depth) by which the market declined.

Impact of Downturns

Looking back at some history, there have been seven major construction market downturns since WWII where the market declined, rebounded, and then went on to continue further growth. As of this writing, we are in the early stages of another market decline, in large part caused by the COVID-19 pandemic. These major market cycles have been roughly ten years apart, with a small number of lesser, not-so-major, dips in between. There could be some debate about this

DOI: 10.1201/9781003229599-17

because there is no agreed-upon measurement of major or minor construction market downturns. For the purposes of this discussion of construction market cycles, a downturn that rebounded quickly enough to support continued annual market growth was not considered major because it had no appreciable impact on the overall construction industry.

Looking at a construction company, the work priced during the healthy economy, prior to a decline, is generally profitable. However, while that work is being completed, the amount of work in design and funding stages is declining, as mentioned earlier. When a construction company's backlog begins to fall off, the pressure is on. It will often feel like a recession regardless of the actual definition of the economy at the time. Prospering in cyclical markets and surviving a downturn begin with recognizing what will happen in the marketplace as soon as the market softens. The result is somewhat predictable and has occurred without fail in every major construction industry down cycle for the last 70 years, starting with a buyer's market.

Buyer's Market

When there are fewer projects in any market, a buyer's market develops, competition intensifies, and prices and potential profits diminish. The ideal in a shrinking market would be for each contractor to accept proportionately less work so that the market share of each business is maintained. However, there is a tendency in our industry to resist any reduction and to compete vigorously for the fewer available projects, driving down prices for everyone. Trying to maintain volume in a declining market is, in effect, an attempt to increase market share, and any increase in market share is always "bought" at a cost.

Construction firms are subjected to both diminished sales and diminished profit, while their cost of doing business, being difficult to reduce, continues. Contractors resisting a reduction in their sales often load up on "cheap work" which increases their risk in an already difficult circumstance. Conversely, cooperating with the market and downsizing to align an organization with market realities are appropriate management practices of the risks imposed through an environment outside of the contractor's control. We cannot control the market, but we can control our response to it. Years of research into every construction market downturn since WWII confirms that the dynamics of each downturn were generally similar. The model developed prior to the last construction market down cycle was live tested during the 2008–2012 recession and accurately predicted each stage of that declining construction market.

The Compression Dynamic

Compression is defined by the authors as the domino effect of the downward pressure applied to construction pricing by large contractors during market contractions. What happens in a downturn is that larger construction companies, desperate to keep their people busy, aggressively go after smaller projects than they normally would. This takes work away from other firms, forcing them to go after smaller projects than they normally would, and so on through midsize firms and finally down to small firms. With their greater capitalization, big contractors are able to use aggressive pricing to take smaller jobs and/or expand geographically by moving into markets they do not normally compete in. The lowest level in this avalanche, the smaller contractors, have nowhere to turn and some are driven out of business.

As an industry, we need to stop believing that it is automatic (almost acceptable) to lose money during a market decline. This happens so regularly to those who refuse to downsize and cooperate with a declining market that we have begun to think it is inevitable. It is not inevitable unless contractors persist in *fighting* declining markets rather than cooperating with them. We have got to stop thinking that we can control the market. The construction market is independent of the construction industry and cannot be influenced by contractors. The construction market is a product of the US economy and results from the needs of parties outside of the construction industry for facilities to be built and their ability and willingness to pay for them. This is the business environment we work in so it is critical that we learn to prepare for, and manage through, the inevitable market cycles.

Market cycles and compression are very real and dangerous realities. Those who don't understand these dynamics and patterns will not recognize them when they occur and will not be prepared to react appropriately. You cannot fight the avalanche once it begins. Some contractors believe that the only way to *fight* compression is to continually lower their prices as larger competitors force their way into their market, which causes losses they should not incur and may not be able to afford. What you can do is refuse to take work at a loss, which means downsizing to cooperate with a declining market. All of us know to get out of the way of an avalanche because defying a landslide can be a lethal choice. From a business perspective, a market downturn is oftentimes no different than an avalanche.

Mini Case Study

The Case: A successful self-performing steel fabricator and erector was doing about four million dollar of work per year when the 2008–2012 construction market downturn hit his company. This young contractor experienced the cancellation of a million dollars of backlog within a 30-day period because his clients on several projects lost their bank funding. The contractor had recently attended one of our seminars where we illustrated that our research models were projecting a deep and lengthy market decline. We also described in detail the concept of "compression" mentioned above. The contractor contacted us to discuss his circumstances, explaining that larger contractors were taking work in his market below his costs just to keep some of their people busy. He said his remaining backlog was uncertain and that he could not see any work on the horizon that he might be able to get for anything but a loss. He said he did not have enough work on hand to make a profit with the size of organization he was trying to maintain. He was not sure how long he could continue to lose money and thought out loud that the time may have already passed.

The Choice: We discussed his circumstances in detail and just by listening to himself he eventually said, "I don't think I can make it and I am afraid this may bankrupt me." He wanted to know what we thought. We told him that we did not know enough about his business or his financial condition to advise him but what we did know is that it seldom makes business sense to put work in place for a loss for any length of time. He asked, "If I quit now do you think when the market recovers I might be able to go back into business?" We told him it should be possible, but to do that he would have to finish the work on hand, satisfy the customers and designers, and avoid bankruptcy which would complicate matters going forward. For his idea to work, he needed to conserve cash, which meant he would have to stop losing money as quickly as possible by cutting all costs except the people and resources required to finish the work on hand, as rapidly as possible. If he could not stop the losses, then reducing them was critical.

The Decision: This business was at the bottom of the compression taking place in its market and was being pushed down with nowhere to maneuver. This contractor initially wanted to stand and fight, take cheap work, and try to ride-it-out, but his calculations and his gut told him he would end up in bankruptcy. He made the monumentally difficult decision to shut down his business immediately, reducing overhead costs to almost zero and keeping just enough resources to complete the work on hand. It took almost a year to complete the projects. The customers and designers appreciated his effort and honesty in dealing with this difficult situation, and just about everyone came to respect his courageous decision. He wanted to keep his equipment, but that would have caused ongoing storage and maintenance costs because the office and shop were rented; so everything was liquidated. When the work on hand was completed, the young contractor took a job with a local general contractor who was glad to have him even after explaining his desire to go back into business in the future if possible. They agreed to a two-year commitment. Being employed in construction, he was able to keep abreast of industry developments, and three years later his market leveled out and began a slow recovery. He parted employment amicably and was able to restart his business. He started small at first primarily brokering work, but a year later was able to move to larger offices with a shop and fabrication yard. By 2019, he was doing $7.5 million a year with 50 employees, a 15-men fabrication department, and record profits.

The Lesson: This case study is not intended to recommend a course of action but instead to describe the "compression" concept in illustrative detail. This process accompanies every major construction market downturn, and construction professionals need to anticipate it, understand it, and act accordingly.

Downsizing as a Defense Mechanism

The compression process highlights the reality that cutting the cost of doing business by downsizing can be an excellent defense (and sometimes the only defense) during a declining market. The path forward is to cut overhead proportionally to the amount of the market decline, enabling the contractor to produce a lessor amount of work at a profit. If there is less work available and a firm attempts to maintain its sales volume, they will have to take the work of their competitors who are trying to do the same thing. Compression causes industry-wide losses which is part of the reason that the construction industry has the second highest business failure rate in the country – a failure rate that accelerates during declining market cycles.

There are only two choices in a downturn.

- **Choice A**: Fighting to capture enough work to maintain sales by any means possible. This is by far the most popular choice but presents huge risk.
- **Choice B**: The less popular choice of downsizing the business to cooperate with (rather than fight) a declining market. This choice has far less risk but is more painful. Downsizing means reducing the cost of doing business (cutting overhead) which includes laying off people. A large portion of overhead expenses for a construction enterprise consist of employee-related costs. Voluntarily cutting back the size of a firm is difficult, even distressing, which is why it is such an unpopular choice.

Chasing work in a declining market while profits are spiraling down makes little sense. One bad job can be devastating because all the other projects have a low margin, so there is not a lot of room to make up for losses. In a growth market there are profits from the other projects to help cover a loss from the occasional bad job. In a declining market, the diminished profits on all projects may not be enough to cover a losing project, and a losing project is an increased possibility because of the aggressive pricing trying to capture the much-needed work. Even during a growing market, construction is a low-profit business with the ever-present possibility of a losing project in both good and bad markets. This exposure is compounded when, motivated by the need for work, a firm decides to go after projects in unfamiliar territory or work of a slightly different type or size than the organization is experienced with, which increases project risk exponentially. There is no statistical data to calculate the likelihood of a losing project during a declining market. However, the fact that the business failure rate in the construction industry increases significantly during downturns suggests a correlation.

Timing of Downsizing

Timing is one of the reasons the business failure rate goes up in a declining market. If a construction enterprise waits too long before reducing overhead, its financial condition weakens. That makes it vulnerable if the downturn drags on and hinders its financial ability to fully participate when the market rebounds. Many have said, *"If I am forced to reduce overhead at some point during a downturn I might as well have cut it sooner."* Unfortunately, many have realized that a little too late. It is disappointing to have spent a lot of money holding onto people only to have to let them go anyway. Many have tried to keep people busy by traveling far and wide, taking work outside the organization's experience and core capabilities, but that often magnified underperformance and in too many cases resulted in losses. During many years of business turnaround work in the construction industry, the authors often heard owners say, *"What choice did I have?"* or *"I had to keep my people busy,"* or *"No one can make money during a downturn."* For those determined to maintain sales or reduce sales as little as possible in a declining market, these statements may be true, and we fully understand and admire the determination to retain employees. It is also human nature to resist giving up hard-earned growth gained at great effort. The problem is the risk and cost, and it often does not pass the logic test.

Consider the following: If you asked a contractor why don't you increase your sales by say 10% or 20% during a favorable market, most would explain that

> You can only grow as fast as the market grows. If the market is not expanding at 10% or 20% that amount of growth would be extremely difficult, and to attempt it I would have to drop my prices, find additional resources, and have to secure them and pay for them.

With this in mind it should also follow that in an unfavorable market, severe adjustments in pricing will be forced into the marketplace and idled excess people and resources will need to be paid for or discontinued. Everyone seems to agree that growth requires an increase in overhead, but they do not seem to agree that the reciprocal is true. The reciprocal is that negative growth (a reduction in sales) requires a reduction in overhead.

In every downturn since WWII, the construction industry suffered reductions in profit margins, some losses, and an increase in the number of business failures. All these occur while the potential for profit still exists if the business adequately adjusts to the new market conditions. The contractor's responsibility is to adapt and manage risks in either case. This includes taking the difficult steps to downsize, including rapid reductions in overhead. Risk control mandates these steps be taken sooner rather than later. The largest of all overhead costs are often employee salaries. While cutting back non-essential costs such as subscriptions, bonuses, travel, and entertainment is appropriate because it signals to employees a new attitude, it rarely amounts to enough to really matter. Unfortunately, a meaningful reduction in overhead oftentimes requires a proportional reduction in management and administrative personnel.

Resistance to Downsizing

We regularly hear from contractors: "I can't operate with 10% or 20% less work. I have a 'drop- dead' volume I have to maintain to be viable." Our response is: "As you grew your business from $5 million and $10 million on your way to $15 million, were you profitable at $5 million and $10 million?" Most contractors were. The point is, if you have to go back to one of those reduced volumes, you need to size and configure the organization to what it looked like when it was profitable at that size. This also includes reducing equipment resources by selling equipment, or mothballing them if sale is not practical, as long as costs associated with them are minimized or eliminated.

There is a tendency to hold on to people and equipment in order to be prepared for when the market returns. However, unless you expect a very short downturn, there is serious risk that resources will be retained at great expense only to let go at a future date; or worse, that the drain on the organization may make it difficult or impossible to finance recovery when the market rebounds, eventually bankrupting the company which then negatively impacts all employees. The market always returns, but accurately predicting the length of a downturn is difficult, and the primary objective should be to profit during a downturn, which ensures survival until the market returns. If the major concern is to be prepared for when the market returns, note that after a downturn there is often ample availability of equipment and people looking for work, so if there is an early rebound, resources will typically be available. Attracting, training, and retaining good people are major issues during the good years, so serious resistance to these suggestions is understandable.

Construction professionals must react to market downturns that they cannot control, with processes that they can control. Unfortunately, the appropriate reactions range from difficult to distasteful. It is difficult but necessary to keep emotions out of business decisions, particularly because there is no minimizing the reality that someone must look people in the eye when they lay them off. It might be well to remember that managing risks to the organization protects the jobs of the employees that remain. Downsizing also provides an opportunity to dismiss underperforming employees, which also results in executing the remaining reduced amount of work with the organization's best people. It is entirely possible to prosper during a down cycle and to maintain financial strength which will be critical when the rebound occurs. Chapter 4 discusses overhead management as a strategy to enable that.

In a country where bigger has often meant better, downsizing is loaded with negative connotations. Downsizing is a deliberate shrinkage of a company's overhead in anticipation of a decline in volume (Figure 13.1). It can be a profitable alternative and an enlightened response to a tough economy.

Figure 13.1 Downsizing is a deliberate shrinkage of a company's overhead in anticipation of a decline in volume.

Managing through Recovery

A study of the economic dynamics of prior construction market rebounds indicates the recovery periods (not just the downturns) are a financial struggle for many contractors. The primary reason is that "growth eats cash," and many firms strain financially during downturns to the extent they have difficulty financing the increasing amount of work during the recovery. The business failure rate of construction enterprises is even worse during the recovery than during the downturn. Contractors need to understand how to approach recovery, which unfortunately can be as challenging as a downturn.

The length and depth of a market slowdown are directly proportional to the level of aggressive pricing that pre-exists and persists as the recovery begins. Construction organizations need to manage cash flow judiciously during recovery to remain financially viable and credit-worthy. They need to develop effective strategies to deal with the significant cash flow demands of growth, and it requires discipline to resist engaging in the "feeding frenzy" for newly available work during early recovery. Comprehensive strategies are necessary to manage risk and to avoid losses, depleted capital, or diminishing equity, weakening the ability to fully participate in the recovery. The contractor's bonding capacity may also be affected by weakened financial conditions, creating additional barriers to fully participating in recovery.

Compounding the problem is the fact that in all of the major construction market downturns in the last 70 years, aggressive bidding persisted during much of the recovery, until the market returned to its prior size. Continuing aggressive bidding compresses margins when they are needed most. Contrast this with healthy growing markets where growth can be financed with profits. Absent the profit potential, the recovery growth will require outside financing, which will be problematic for financially weakened firms. Combine this with the uncertainty that the banking industry may not be ready to immediately reengage in construction lending, which leads to the "perfect storm" for some companies.

This exposure is compounded by the fact that after a sustained downturn, inflation is common, caused by labor and material shortages and material cost escalation. Some laid-off construction tradespeople and managers who have vacated the industry during a protracted downturn do not return. Material manufactures and suppliers that have been forced to cut capacity and have been unable to raise prices need time to regain the capacity and required investment, which often causes material shortages that lead to rapid price increases. Prospering in cyclical markets and surviving both a recession and recovery in the construction industry begin with recognizing what is happening in the marketplace.

Oddly enough, one favorable aspect of downsizing is improvement in cash flow. If a business is doing less work each month compared to the prior month, they will be collecting payment for the greater amount of work completed the prior month, but spending less doing a lesser amount of work in the following month. Over time cash flow can improve considerably. The opposite occurs during growth: a firm doing more work each month is collecting payment for a lesser amount of work in the prior month and incurring increased cost to produce a greater amount of work the next month. This has nothing do with profitability and has the same results regardless of whether the work is profitable or unprofitable.

One key consideration to keep an eye on throughout the recovery (and beyond) is debt and its associated increase in risk. Current construction industry margins are tight to the extent that one bad job can consume the profits from two or three other profitable jobs. A careful study of the debt structure of our industry shows we're trying to borrow our future. We contend that the debt to sales ratio for construction firms should be relatively constant, and changes to it should be intentional and justified. If a contractor is doing $10 million in sales with $1 million of debt, logic suggests when they grow to $20 million in sales they should be around $2 million in debt. However, our experience is that the industry's debt to sales ratio has been going up over the last 50 years, particularly for equipment-intensive contractors. This indicates an increase in risk and can threaten the stability of the industry if the trend continues.

Alternative Approach to a Cyclical Market

Contractors can and should prosper in a cyclical market. We know in advance that about every decade or so, the construction market will turn down for a year or more, so we must plan for it. It takes a while to become good planners and to trust your forecasting, but in predicting the market it is wise to guard against excessive optimism. Those who have embraced the "Flexible Overhead"[1] concept and processes and introduced them into their companies have little concern about downturns as they can reduce overhead within a week and without disrupting their core team. The successful contractor of the future can enjoy success in both growing markets and declining markets (Figure 13.2). They will easily downsize to deal with even brief market declines or be able to increase the size of their organization to take advantage of short-term opportunities that require rapid expansion. Every industry evolves over time, and construction is no exception. It is time to alter the business model to flex with the market and to gain the skill sets to prosper through the reality of periodic construction business cycles.[2]

Figure 13.2 The successful contractor of the future can enjoy success in both growing markets and declining markets.

Review Questions. Check All That Apply.

1 What is the lag time of the construction market?
   a 6 to 12 months
   b 6 to 18 months
   c 12 to 18 months
   d 12 to 24 months

2 Which of these are not listed by the authors as choices in downsizing?
   a Fighting to capture enough work
   b Use of your line of credit
   c Downsizing the business
   d None of the above

3 What do the authors list as reasons given by contractors for not downsizing?
   a What choice did I have
   b I had to keep my people busy
   c No one can make money during a downturn
   d All of the above

4 After a sustained downturn, inflation is common, caused by:
   a Labor shortage
   b Material shortage
   c Material cost escalation
   d All of the above

5 According to the authors, how long are construction market cycles?
   a They vary widely
   b The market cycles every 15 years
   c The market cycles 8 years
   d None of the above

Critical Thinking and Discussion Questions

1 Discuss the compression dynamic.
2 Explain downsizing as a defense mechanism.
3 Discuss the resistance to downsizing described in this chapter.
4 Discuss the authors' recommendation about managing through recovery.
5 Explain the section on the alternative approach to a cyclical market.

Notes

1 *Flexible overhead is covered in Chapter 4.*
2 *Business planning is covered in Chapter 3.*

14 Construction Industry Consolidation

Industry consolidation is characterized by the amalgamation of smaller organizations into larger ones, resulting in fewer but more powerful industry participants (Figure 14.1). Consolidation activity usually signifies an industry is in a mature phase of its life cycle, as established businesses acquire competitors in an effort to strengthen their market position. The rationale for consolidation includes greater negotiating power with suppliers and customers, access to new technology and practices, and expansion of markets and product lines.

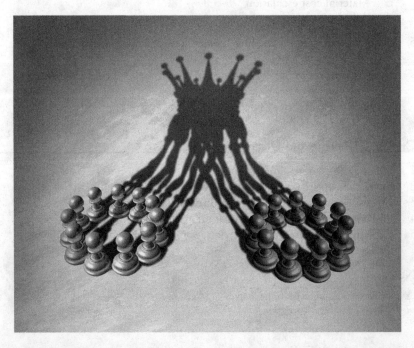

Figure 14.1 Industry consolidation is characterized by the amalgamation of smaller organizations into larger ones.

DOI: 10.1201/9781003229599-18

Consolidation in the Construction Industry

The construction industry in the US has been leaning in the direction of consolidation for a long time but began to get traction after the first decade of this century. Around year 2010, consolidation of the construction industry was measurable and heading in the direction of 25% fewer enterprises, which is a generally accepted definition of industry consolidation. In the September 18, 2000 issue of *Engineering News-Record* (ENR), one of the authors of this book projected that the number of existing construction enterprises would be reduced by more than 20% by 2020, which is about where the industry is at the time of this writing. When large construction enterprises grow at a rate faster than the remaining firms in the industry, it is at the expense of the midsize and smaller organizations who are left with a smaller portion of the market. Big construction companies are getting bigger, while midsize organizations struggle with proportionately less work.

This is the case in various market sectors. For example, one of the authors and his team also analyzed this phenomenon specifically for design-build firms[1] using revenue data from the top 100 design-build firms. The authors divided the firms into statistical subsets (also called clusters) and analyzed the growth rates of each subset compared with the average of the top 100 design-build firms as a whole. They found that the top 17 firms (by revenue) actually have 70% of the revenues of the top 100 firms. Moreover, looking at these top 100 firms between 2003 and 2013, revenues for the top third of these firms grew at a rate double that of the bottom two-thirds, significantly increasing their market share.

From 2008 to 2012, the work put in place by the entire construction industry was reduced by 19.6%. However, over the same period, the work produced by the ENR-listed top 100 construction firms shrunk by only 14.5%, which means the largest 100 construction firms captured 5.1% more of the work than the remaining construction firms in the US. The remaining construction firms got disproportionately smaller, while the 100 largest businesses captured a greater share of the shrinking market, and that trend continues today. When this occurred in other industries, it led to consolidation, with small and midsize firms merging to become larger firms, or being absorbed into large firms.

Comparison with Other Industries

To better understand the issues, let's look at other industries that have consolidated. Before WWII, the majority of the food and groceries consumed in the US came through small "mom-and-pop" grocery stores. In 1950 when one of the first grocery store chains opened, all the "mom-and-pop" grocers said in unison: *"They'll never make it. A person needs to know their grocer personally."* Today most groceries come through huge chain stores. The same goes with privately owned gas stations (formerly referred to as filling stations) having been replaced with franchises or oil company-owned stations. This is even the case for farming, where most small and independent farms in the US have been consolidated into large corporate operations. Maybe construction industry customers will be more

loyal, but other industries became commodities with minimal resistance from customers as these industries consolidated.

Historically the construction industry had been an exception because the
amount of construction put in place by some large and publicly traded construction corporations had been minor compared to that put in place by more than a
million separate and independent small to midsize contractors. However, in the
last couple of decades this has slowly started to change, with larger firms getting
larger, and the process is accelerating, which is a pretty good sign of an industry
consolidating. Some are saying, "it can't happen with construction" ; however, it is
already happening even in the architectural and engineering side of our industry
where consolidation is progressing rapidly.

What Changed?

In the past, success in the construction industry demanded a rare combination of
talents, including the abilities to muster resources; put an accurate price on the
work in advance; manage labor, subcontractors, and vendors through a long and
arduous process; and tolerate a high degree of risk. Many projects were built with
a backhoe and a box full of tools. In the old days, some used to say, "all you need
to be a contractor is a pickup truck, a cast-iron stomach, a forgiving spouse, and
a bad temper." Contractors of this breed were successful as long as their methods
of bidding and costs for individual items of work were closely guarded secrets, and
as long as their production of complex projects remained a mystery. But all of that
has changed. New technologies and wide access to information have demystified
the processes of estimating, organizing, and producing the work, resulting in construction being perceived as a commodity. This led to a belief that there is little
difference between construction enterprises, which resulted in less concern over
which contractor is selected. Note that this belief is slowly starting to change
again, with the emergence of alternative delivery methods such as CMAR and
design-build DB where contractors are selected in large part based on their qualifications and not just their price. But that's still the exception, not the norm.

But again the most important reason is arguably construction being perceived
as a commodity, in a world where processes are evolving and efficiencies are improving. Here's a simple example: not that many years ago, wooden forms for
concrete were built from scratch to custom design even for complex projects.
Competition for that type of work was limited to contractors with craftsmen who
knew how to do that. But since the process has been revolutionized by patent
form systems, it now requires much less skill to build forms. Standardized processes and lower costs invite new competition, and some small and midsize construction contractors resist change or lack the resources or motivation to change.
Moreover, technology is improving standardization, efficiency, and productivity.
It is also streamlining planning, scheduling, and production to the point that
construction is no longer the custom product or service that it was.

Fifty years ago, construction was a custom effort, with double-digit profit margins compensating for inherent inefficiencies. Margins have slipped into the mid

and low single digits, and even lower for larger contractors. Margins have been "structurally" altered and are unlikely to grow substantially in a hard-bid market. Construction has become increasingly standardized, leading to efficiency and productivity becoming major differentiators among competitors. Economies of scale and homogenous processes reduce cost and schedule where larger firms have the advantage of greater resources. Many small and midsize construction companies lack the means to reduce cost and schedule to the extent today's customers are demanding.

Consequences

The reduction of margins over time makes it harder for small and midsized businesses to invest in increased efficiencies. Larger companies suffer the same low margins, but their size enables them to invest in production improvements because they can better afford the upfront costs involved, to satisfy the shifting demands of owners. Economies of scale make larger organizations best suited for survival. Because continuous improvement is a necessary cost of doing business, consolidation will become critical to survival for some contractors.

Some small contractors will fare better than midsized firms because there is usually a place for niche players. Specialty contractors can earn higher margins, but that may be impacted as consolidation continues. Small and midsized firms may attempt to match larger-scale efficiencies through cooperation, trade associations, or seeking to work together, but their window of opportunity is brief. Profitable well-managed midsize organizations have the options of becoming part of larger organizations or finding and serving niche markets.

Many contractors will resist losing independence through acquisitions and mergers, but this reaction has not been very successful in other industries that have consolidated – unaffected by the resistance of the participants. The talents, aptitudes, and temperament of the traditional successful contractor generally fit the mold of the headstrong entrepreneur who loves to lead (not to follow), which partly explains why the industry has remained fractional for so long.

Consolidation leads to standardization, which demands efficiency. Therefore, the future leaders of the construction industry will likely be efficient, well led, well trained, large regional and national organizations. Consolidation does not just mean a few roll-ups, but rather a serious reduction in the number of US construction companies. If construction follows the pattern of other industries that have consolidated, more than 100,000 small and midsize construction enterprises will be absorbed or will disappear in this decade. Because everything moves fast in the information age, the timing is unpredictable. Travel agencies, for example, disappeared almost overnight. The more quickly consolidation occurs, the more difficult it will be on practitioners. Although the same construction professionals will continue to manage and perform the work, leadership will shift toward larger firms. The structural changes will create uncertainty, which is generally unpopular. This subject is not easy to digest, and we suspect will be rejected by many.

In a recent turnaround effort in the Midwest, we tried to convince an underperforming general contractor that his market had changed dramatically. Several of his steady customers had been acquired by national firms that were now giving their "local" work to regional and national contractors brought in by the new regional and national owners of their former customers. At the same time, several of his competitors had grown substantially or had merged with larger firms. What he needed to do was get bigger or smaller, but with his limited resources, getting bigger was not an option. He told us: *"I didn't put my entire life into building this company to give up now."* However, his refusal to react appropriately cost him his business. In our work we counsel construction firms about performance improvements and profitability enhancements, but it is difficult to advise about consolidation because most owners do not want to hear it.

Construction Industry Changes that Support Consolidation

There are a number of changes in the construction industry that impact risk and, some would say, "take some of the fun out of the business." These changes may add to the willingness of contractors to sell, merge, or otherwise leave their businesses:

- Reduced margins make the business less attractive
- The return on investment (ROI) is not as appealing as it was in the past
- Inheritance tax is difficult on successors
- Lack of family successors
- Projects becoming more complex over time
- Increasing demand for speed
- This already high-risk business continues to get riskier

The above realities often cause immediate family members to be less interested in the long hours and hard work required when taking over a family-owned construction business. This will make a number of current construction entities "last generation."

Drivers of Consolidation

As mentioned earlier, economies of scale provide larger enterprises price and competitive advantages not available to small or midsize organizations. These include:

- Buying materials in bulk
- Attracting subcontractors to larger jobs and the potential for repeat business
- High and consistent use of equipment
- Efficient use of labor, technology, capital, and management
- Cost-effective assembly of standardized parts and prefabrication
- Standardization of construction practices
- Separate experts in design, purchasing, production, quality control, safety
- Resources to continually explore the latest technologies, and more.

The large firm advantage is cumulative and provides buying power, cost savings, and leverage that cannot easily be matched by small or midsize companies. The larger firm can build for less and sell for a reduced amount. The threshold of the profit percentage it takes to satisfy a large firm's single or multiple owners or stockholders is often less than the profit percentage required to support a family-owned business or satisfy the independent small or midsize contractor.

Buyers of construction services are also changing. The average size of buyers is increasing as corporate America grows and consolidates. The size and number of multinational companies is growing. Government agencies at all levels are often aggregating projects for more efficient management. When buyers' preferences shift, project delivery does too, new providers emerge, and the potential for consolidation increases.

Impact of Consolidation

Smaller contractors will be least affected by consolidation because most are, or can become, niche firms. It may seem counterintuitive, but the small contractor has the defense option of getting smaller if need be. This is a luxury the midsize contractor often does not have, so the midsize contractor must recognize when consolidation will affect them. Larger firms are drawn to growing areas, and a primary entry method to a new market is to purchase an established local midsize firm with the intention of growing it. The standard practice is to select what the large firm considers the strongest, well-managed contractor in the location and pay a fair price, occasionally overpaying for a quality firm that may be a good fit. Contrary to some opinions, enlightened large firms are not attracted to the cheap price for a weak firm or one that may be struggling with financial difficulty. Experienced large firms entering a new market usually acquire one of the top three quality, reputable midsize companies in the area. The typical reasoning is that they don't want to tarnish their own reputation and prefer to enter the market with a strong partner, ideally with some profitable backlog. Next, we illustrate this concept with two case studies.

Case Study #1

A very successful second-generation non-family midsize contractor in a growing city was approached by a well-respected national firm expressing an interest in acquiring them. The professional owners of the midsize firm were surprised because they had never considered selling their business and asked why they were being approached. The representative of the national firm advised that they were going to expand their operation into the area and that their research indicated the case study firm was the best in the area. After disclosing that they were not necessarily for sale but would be interested to learn more details, an all-day meeting was scheduled.

The partners met with senior executives of the national firm and were overwhelmed with the purchase amount they were offered. At lunch time it was suggested that they dine separately so the partners could have a discussion among themselves. The partners were genuinely flattered with a commitment that they would still be running the operation and by what they considered a very generous purchase price but decided they were not interested in selling. They advised accordingly and the meeting ended cordially. The partners did not give it another thought until three months later when it was announced that the national firm had acquired the partner's direct competitor that was somewhat larger than the partner's firm and had always been a serious challenger for the best work in the area.

Within months it became apparent that the newly acquired firm had home office marching orders to increase their market share considerably, and the funding to accomplish it. Shortly thereafter it was learned that the acquired firm planned to double in size, which explained why they were so aggressive in pricing the work. Within two years they were able to capture the majority of the larger projects that the case study firm's partners were counting on. During that time, the acquired company was awarded two projects so large that they were outside the partner's range because they did not have the bonding capacity to even compete. Despite a growing healthy market, the midsize case study firm was unable to capture enough work to maintain their size. In the first year after their competitor's acquisition the firm was involuntarily downsized by 15%, then by another 10% the following year. Over the same length of time, the competitor, now owned by the national firm, doubled in size in the rapidly growing market and increased their market share. They were positioned to maintain that position and grow from there. That was later determined to be the national firm's plan from the start. That business plan originally had the partners' name on it but when they declined it went to their competitor.

The lesson: The partners based their decision on their personal wants and needs because they owned the company. However, they did not consider any ramifications of declining because they did not think there were any. Construction professionals need to see beyond their internal perspective and understand what the initial inquiry might mean to the firm, the area construction market, and to the local construction community. One thing that should have been considered is that the national firm was coming into this area no matter what (which they were told). That fact changed the circumstance and question from: "Do we want to be acquired?" to: "How will this affect our competitive position, our area's construction market, and the local construction community?"

This is a very different circumstance that the partners were not aware of. They decided later, that had they known that their company would be so negatively impacted by not being acquired, that they would have agreed to the offer. They are now convinced they made the wrong decision. The lesson is that consolidation is occurring in the construction industry and it will impact those firms that consolidate and those that do not because it alters the competitive balance and the construction market. An exception might be a midsize company in a rural area where large projects are rare and there is limited competition because that circumstance does not readily attract large or national firms. However, that may change over time, so awareness of consolidation is critical for all senior construction professionals.

We are guessing that some will disagree with the lessons learned from this case study, saying the partners own the business and can do what they want with it, or they do not need to work for someone if they don't care to. On the face of it that sounds correct, but it suggests that top management should consider what they want or need or what is best for them personally in their decisions, which is not always correct. Senior management of a corporate entity, whether owners or not, have a primary responsibility to do what is best for the corporation – the same responsibility that boards of directors have. That is what stockholders expect and deserve, and the fact that the stockholders and top management are one and the same does not change that.

Case Study #2

This well-respected contractor was the largest contractor by far in his rural state with no competitors of his size. As such, he captured all the large projects in the state. There were not many large projects and they are relatively not that large compared to much of the country, but he got them all anyway. He also got a good share of the midsize projects. Then things changed.

There was a large out-of-state manufacturing company that had locations in the contractor's state, and he built all their major projects. During the 2008–2012 recession, his largest-ever project was canceled by this manufacturing owner just as it was getting started. About four years later, a national construction firm contacted the contractor asking if he would joint-venture with them on the very same project which was now two and a half times bigger. The contractor immediately called who he thought was his client, and was told that because of the new size of the project he no longer qualified financially for the project by their standards and that they had recommended him to the national firm as a potential local joint-venture partner. He was naturally upset by the circumstances and refused to joint-venture because he thought it would be "helping" the national contractor build what he considered "his job."

Because there was no other contractor in the state large enough or experienced enough to become a local joint-venture partner, the national firm moved an entire field management team to the site to build the project. Because of their presence in the state for the year and a half it took to build the project and through the national contractor's many connections with national buyers of construction services, they were given a number of other large projects in the state. Those projects normally would have been the work the case study firm would have gotten.

Well before the original project was completed, the national contractor realized the construction market in the state was growing and that there was only one large contractor in the state who seemed to have a monopoly on the market. The national contractor decided to establish a branch and expand into the state. They rented offices and yard space and became our case study contractor's "worse nightmare" (*his words, not ours*). The national contractor was now the biggest contractor in the state and their branch office began to grow from the day they opened the doors. Our case study contractor captured less and less work, was reduced to 60% of his original size, and never regained his position as the premier contractor in his area. He joined the ranks of the midsize contractors that previously couldn't compete with him.

The lesson: The case study contractor later said to us,

> I should have joint-ventured with the national firm. I didn't because I let my emotions get the better of me. If I had worked with them they never would have had such a strong presence here in my state and certainly they would never have brought so many senior people here because, as the local partner, I was to provide the management for the project. We would have finished the job and they would have left without ever knowing about the market that I had all to myself.

It is a form of consolidation when the big get bigger and someone else gets smaller. When this contractor explained what happened, he said

> I had never even heard the phrase 'construction industry consolidation' before. I wish I did because I had the opportunity to prevent this national contractor from getting a foothold in my market. I am paying for it now and my family also will, because the family business will never be the same.

Conclusion

Consolidation is here to stay. Some companies will pass up the chance to become more efficient by not merging with others or being acquired. To be the best requires continuous education and training, but some contractors tell us, *"I can't afford to upgrade; we are always too busy for training; etc."* Low margins restrict resources, making it harder for small and midsized businesses to invest in increased efficiencies. Larger companies suffer the same low margins, but their size enables them to invest in production improvements, to satisfy the shifting demands of owners.

Ironically, some small contractors will fare better in consolidation than some midsized firms, because there will always be a place for niche players. Specialty contractors will enjoy relatively higher margins until consolidation moves ahead at full throttle. Small and midsized firms may be able to accomplish large-scale efficiencies through cooperation, but their window of opportunity may not be open forever.

Review Questions. Check All That Apply.

1   The generally accepted definition of industry consolidation is:
    a   15% fewer enterprises
    b   20% fewer enterprises
    c   25% fewer enterprises
    d   30% fewer enterprises

2   Fifty years ago, construction was a custom effort with double-digit profit margins compensating for:
    a   The high level of risk
    b   Inherent inefficiencies
    c   The degree of difficult
    d   The need for construction services

3   Which of these is a cause of construction industry consolidation?
    a   The degree of difficulty in building
    b   The attractive profit opportunities
    c   The increasing need for infrastructure
    d   None of the above

4   Which size construction enterprise do the authors say will be least affected by consolidation:
    a   Small firms
    b   Midsize firms
    c   Large firms
    d   Multinational firms

5   What is the authors' conclusion about consolidation in the construction industry?
    a   It will slow down
    b   It will never gain traction
    c   It is here to stay
    d   It can't be predicted

Critical Thinking and Discussion Questions

1   Explain the authors' claim that some small contractors will fare better in consolidation than some midsized firms.
2   Discuss what change in the construction industry led to consolidation.
3   What have been the impacts of consolidation on the construction industry?
4   In your opinion, which are the more impactful drivers of consolidation listed?
5   Discuss how you think consolidation will impact the construction industry in the future.

Note

1 Vashani, H., Sullivan, J., and El Asmar, M. (2016). "DB 2020: Analyzing and Forecasting Design-Build Market Trends," *Journal of Construction Engineering and Management*, Vol 142(6), p. 04016008.

15 The Construction Company Life Cycle

Construction companies go through stages as they grow from startup to maturity. To evaluate and manage a closely held construction enterprise, it is necessary to understand which stage of growth the company is in. An extensive study of closely held construction firms reveals that there are five distinct stages of growth: startup, survival, success, growth, and maturity. It is critical to understand what stage a firm is in because the elements needed for success in one stage can spell disaster in another.

A construction company generally remains in the startup stage for five to ten years and then moves into the survival stage or ceases to exist, as many startups do. The business failure rate during startup is extreme, with most studies stating that from 50% to 70% of startups fail. If a startup survives, it moves to the survival stage, which in the best case can be passed through in a few years, while on the other extreme an underperforming construction company may remain in the survival stage indefinitely. The failure rate in survival is high but less than in startup. A company that makes it through survival stage graduates to the success stage. The key to moving on to the success stage is a significant reduction of debt. Therefore, it is not possible to move from survival (where cash is scarce) directly to the growth stage. The reason is that growth requires funding, and cash is earned and accumulated in the success stage, which precedes the growth stage.

The success stage can be a launching pad for growth, or an organization can choose to remain in the success stage indefinitely. On rare occasions a company can pass from its success stage to the maturity stage without entering the growth stage. If a company remains in success for many years, it may slowly grow to a size where it has sophisticated systems and planning and becomes well-established in its marketplace that it enters the maturity stage. It may not go through the growth stage if the company never grows at a rate greater than 15% per year (plus inflation) and does not increase debt during its moderate growth rate. A construction company seldom reaches the maturity stage before the second generation, and it usually takes even longer than that.

It should be understood that a company cannot choose or decide what stage it is currently in. The stages are defined, and a company fits into one because of the company's current circumstances and conditions. A firm claiming to be in a stage does not make it so. This is particularly significant when a firm in growth or

DOI: 10.1201/9781003229599-19

mature stages has financial problems that move the firm back to the survival stage and it does not realize it or refuses to acknowledge it.

Stages of Growth

The following is a discussion of each of five stages of growth that a construction company may be in. They are most commonly passed through from startup, to survival, to success, to growth, to maturity, but regression is also possible and not uncommon, particularly during times of financial difficulty (Figure 15.1).

Stage 1: Startup

A startup construction company is generally run by a contractor/entrepreneur who does everything: bids the work, sells the work, supervises employees, and provides working capital. The startup lacks any assurance that it will survive the next fiscal year. This is not to say that it is in trouble or that it is not making a profit. It simply has no track record yet. It has not proven its capability to get and produce work at a profit on a consistent enough basis to satisfy its customers,

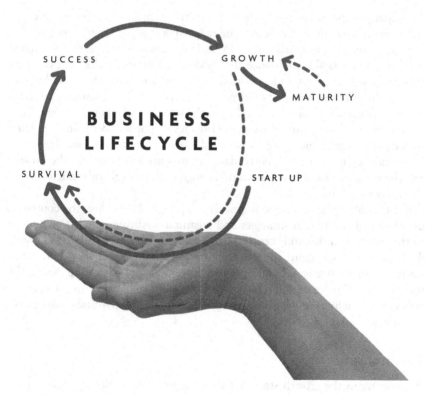

Figure 15.1 The five stages of growth.

potential credit grantors, or itself. A company may stay in the startup stage for ten years or more, though some firms progress more quickly. Optimism, perseverance, and endurance are key ingredients during these years.

Stage 2: Survival

In the survival stage, the company is stable, and the founder and credit grantors believe that the enterprise is more or less permanent. The company has discovered its preferred type of work or found a business niche. It is still a one-person show (or perhaps just a couple of key individuals), but the primary concern is no longer to get from one month to the next, because it has grown to a sufficient size to generate enough revenue to solidify the organization. Cash flow is the bane of the contractor's existence in the survival stage. The contractor can borrow money because the company is relatively stable, but usually learns quickly that growth consumes cash and that more borrowing is necessary to grow. There is little organizational structure or control processes. Managers follow the very specific instructions of the founder.

Stage 3: Success

A company in the success stage generates positive cash flow from its operations. If the company owes money, debt is being reduced or it may even be debt-free. The company is growing modestly (under 15% sale increases year-to-year) or its volume fluctuates, but it is able to finance this modest growth internally. The company is large enough to maintain its position in the marketplace and earn at least average profit margins compared to its peers. The risk of failure is minimal, and the personal assets of the owner are no longer the sole basis of credit. Managers have responsibilities and are functioning at a high level. The founder is still in charge but may have some time to pursue personal goals if he or she chooses. There may be a tendency to overspend, particularly on benefits and bonuses. The greatest risk the company faces is maintaining enough cash to deal with tough times if those were to occur.

If the company has continuous growth at a pace of 15% or less, the contractor usually remains active in strategies and organizational issues, and is committed to the established demands on their time and energy. Owner time and energy demands vary from organization to organization but are usually less than during startup or survival stages. Employees may have difficulty advancing during this stage because the organization has leveled off and is stable. At this stage, profits are made by maintaining business-as-usual and managing risk and causing cash, revenue, and equity to grow.

Stage 4: Growth

A company in the growth stage is increasing its annual sales by 15% or more (on top of inflation). It is usually profitable, but often at a lesser rate than when it was in the success stage. Depending on the rate of growth there is a risk of becoming unprofitable, at which point the firm would return to the survival stage (regardless of whether that backward move is acknowledged by management). The major issue in growth is financing that growth or expansion. Debt is almost always increasing because growth consumes cash, working capital is strained, and retainages increase. Working capital can improve if growth stops at a plateau. Cash flow is often a problem in the growth stage. This can be discouraging for the entrepreneur that has spent time in the success stage where cash was plentiful. There will be an increase in debt to equity ratio that can impact bonding and banking relationships. The ability to delegate authority becomes critical, because in this stage there are more people to manage. This is a difficult transition for many entrepreneurs, particularly those who have not had to delegate in the past or have difficulty in delegating at all. The contractor may place strict controls on those delegated to, which limits the potential to take the load off the entrepreneur, who is often swamped during his or her firm's growth stage.

The "growth stage contractor" must begin to decentralize. Existing systems and procedures become strained; replacement systems often become overloaded before they are operational. Long-range planning is critical and must include the participation of key managers, but time for planning is usually in short supply. The company is still dominated by the owner(s), but other key people have an important role in management, operations, and hopefully decision-making. The growth stage is a difficult one. Demands on owner time and energy are high. Having enjoyed the cash flow and reasonable time demands of the success stage renders tight cash flow and high time and energy demands less palatable.

Stage 5: Maturity

A construction company seldom reaches the maturity stage during the first generation of leadership. It is usually the second generation or even later. The mature company has competent, self-reliant management and is large enough to dominate a market. Debt is modest, and the company's course is charted through formal strategic planning. The management structure is clearly defined and most likely decentralized and departmentalized. Separate profit centers, departments, divisions, or operational units provide necessary opportunity and training ground for future managers of the company.

If a company in the mature stage begins to grow at a rate greater than 15%, it shifts back to the growth stage, and the systems and controls relied upon in maturity will stifle expansion efforts. If that growth creates losses or substantial debt, the company may reenter survival stage, regardless of its size and reputation, and regardless of whether management recognizes it or not. The mature company's biggest challenge is to control its future without losing its intensity. Mature companies are not always the most profitable contractors in their market. Companies managed for long periods of time under effective systems and controls do not need, and often may not have, very strong leadership, or may lose determination and enthusiasm. However, that is not always the case.

Key Considerations in Various Stages of Growth

As a construction company matures, there are various considerations that management needs to account for to lead the company through its current stage and keep it profitable. These include market factors and business resources, as discussed next.

Depending on the life cycle stage the company is in, the market factors that competitors face can affect each company differently. A favorable construction market may encourage growth; however, growth over 15% per year can shift a company from one stage to another. If management doesn't sense the shift, it may be making incorrect assumptions to inform its management actions and decision-making. For instance, a successful company in rapid growth may be forced to increase debt to finance the expansion, and as a result the firm will shift from the success stage to the survival stage. If management does not realize that the shift has happened, it could be disastrous because a company in the survival stage must be managed differently than a company in the success or growth stage.

Organizations shifting from one stage to another without a change in resource allocation and a recognition of the necessary shifts in management practices often run into difficulty. Even modest or partial shifts form one stage to another create challenges if the change is not recognized and managed accordingly, because an organization in transition may be in two stages at once, adding a layer of complexity to managing the business. This may explain some industry confusion, such as why some contractors need to work long hours each day to keep their business operating while others play golf and their business goes along on its own; and why some organizations manage without any formal systems or strategic planning while others rely heavily on planning and systems for their success. For example, long-range planning is essential to a company in its mature stage, but can be viewed as a distraction for a startup where the contractor is living day-to-day and needs all their time to operate the business as opposed to plan for the future.

Every construction organization relies on business resources for success; however, the demand for particular resources can vary significantly depending on which stage of growth a construction company is in. Primary business resources for management to consider are:

- Owner's ability
- Business connections
- Cash flow
- Matching business and personal goals
- Quality of and diversity of employees
- Owner's capacity to delegate
- Strategic planning
- Systems and controls

Next is a discussion of each of the business resources that a construction company must have in order to be successful, including an illustration depicting the importance of each during different stages, as shown in Figure 15.2. The figure indicates the relative degree of importance on the vertical axis, across the various stages of growth shown on the horizontal axis.

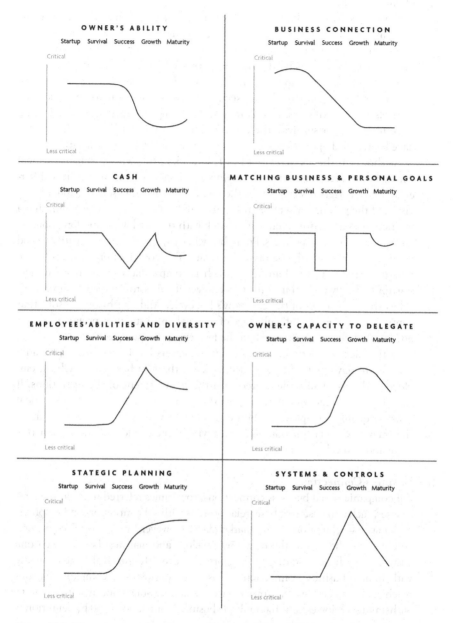

Figure 15.2 The resources that a construction company must have to be successful across its life cycle stages.

A Owner's Ability

In the early stages of a business, the owner's ability to do whatever the company specializes in may be the only resource the company has. If the owner's ability is sufficient, the contractor can sustain a startup company. It becomes less important in the later stages where qualified managers and skilled staff are necessary for survival.

A mature construction company can exist for generations absent a strong owner's ability, because qualified executives, appropriate systems, and adequate controls manage the business. However, if a company shifts from the mature stage to growth or survival, the owner's ability and leadership, which may not have been called upon for some time, become critical to success, again.

As shown in the top left corner of Figure 15.2, the owner's ability is critical during both startup and survival. Having one or more owners makes little difference; what matters instead is that the founder(s) have a reasonable and aligned vision of the product or service they intend to "sell," is/are capable of running a business, and is/are driven enough to stick with the hard work and long hours it takes to launch a new business. Profitability in early stages is rare, so patience and perseverance are tested, potentially for years. The construction business failure rate at startup is worse than 50%, which is comparable to most new business startups. The owner's ability to do whatever it takes to succeed is most critical in the early stages of the company's life cycle. And for obvious reasons that continues through the difficult stage of survival, until stability, profitability, and positive cash flow are achieved and the firm enters the success stage.

In the success stage, the owner's ability matters but is less critical, because in order to have arrived to the success stage, the firm has some qualified employees who assist to some degree with the management of the operations. If that assistance reaches a high degree, the owner's ability becomes less critical because qualified employees' abilities combined with systems and procedures that may have been introduced are carrying some or most of the load at this stage and beyond.

B Business Connections

All companies need business connections, sometimes referred to as business resources, such as a customer base, relationships with subcontractors and suppliers, and a strong reputation in the marketplace. However, the importance of such connections changes as the company develops and matures. Loss of even one customer can be devastating to the startup or even the survival stage company with limited business connections. This is in contrast to a company that has reached the success stage, growth stage, or mature stage, and is better able to withstand such losses, as it has multiple business connections and by definition is established and has some reputation in the marketplace.

Business connections at a startup construction firm may be few or nonexistent, so making positive connections with subcontractors, suppliers, and customers is critical because without them the business will not continue. Due to their size and short time in the market, startup construction firms initially offer limited business to subcontractors and suppliers. Therefore, they may need to put additional effort to get the attention and interest of those subcontractors and suppliers to do business with their new firm, and hopefully to extend credit in some circumstances. Some potential customers may avoid doing business with a startup, and the new firm is obviously unknown to most of the customers that can use their product or service, so those connections are clearly critical. A banking relationship is critical because financial support is usually required at startup. Moreover, if the work is required to be bonded, surety connections are in that same category. Basically, a new firm is entering an existing business community that they need to function in, but the business community probably does not need the new firm as much as the new firm needs them. As the company matures, it becomes a member of the construction business community and these connections develop and therefore become less of a risk. There are many other business connections not mentioned here that are significant, but typically less critical.

C Cash Flow

Startups often cannot get enough cash and credit. In survival, the challenge is to generate or borrow enough cash to break-even and cover the cost of replacement of equipment, and to eventually earn enough to finance growth to a size that provides an economic return on assets and effort. In the success stage, enough cash is earned to reduce or eliminate debt, but overspending is sometimes a problem. In the growth stage, cash is short again and there is a tendency to over-borrow. In the mature company, cash flow meets all business needs and there may be some modest debt in the capital structure. Managing cash flow is easier if there is an understanding and anticipation of the different demands in the various stages of growth of the construction company

As shown in Figure 15.2, the need for cash on the chart starts high on the critical scale, as most startup construction companies need more cash than originally planned, and this criticality continues into the survival stage. A startup makes it through survival when cash flow begins to improve, and if the improvement continues, the firm enters the success stage. Cash is always a problem at startups because there is almost never enough cash earned initially to cover costs.

Once the organization gets to profitability, negative cash flow actually persists because work is put in place before it can be invoiced, and customers do not pay instantly. When or if cash flow gets positive, the firm is still out of all the money it expended to get to that point, so it remains in the red as far as *cash-in* minus *cash-out* is concerned. Startup business owners should know that even with initial success they should plan on generally having negative cash flow for a minimum of two years, with the average being closer to three years. In our experience, for a new construction business to recover all costs from startup forward, and achieving break-even, typically takes two to four years. Those that do not break-even in four to five years usually stop trying.

D Matching Business and Personal Goals

As harsh as it sounds, the startup contractor usually has to put personal goals aside except for the desire to succeed in business. When a contractor is unwilling to do that, or when their personal and business goals are out of alignment, there is reduced focus on the business and reduced potential for success, particularly in the startup, survival, and growth stages.

Figure 15.2 shows that matching a contractor's business and personal goals is critical at the startup stage and remains critical through the survival stage. If sacrificing personal goals becomes too much for the business owner, and he or she decides it is not worth it, at any time during startup or survival, they either quit or withdraw energy from the effort, and the company typically fails.

When the company enters the success stage, the matching of goals is less critical. The time demand on the business owner is reduced, making the matching of business and personal goals relatively less important to the company's success. The reason is the company is stable with positive cash flow and moving out of debt, so there is some time available for personal interests. Not every entrepreneur takes the time, but either way it typically won't impact the company as much.

In the success stage there is some freedom, but not if the company begins to grow at a rate in excess of 15%, as this may signify a shift into growth mode. Note that *matching business and personal goals* immediately returns to highly critical if the firm enters the growth stage because of the need for leadership attention during growth.

In the mature company, matching owner business and personal goals is generally less critical, but again if a decision is made to expand, the company principals must be prepared to set personal goals aside because of the time demands of the growth stage. If a firm spends a long time in the mature stage, the organization sometimes relaxes too much, and the curve turns up again which signifies leadership attention is required to renew the firm's energy.

E   Caliber and Diversity of Employees

From startup through success, contractors typically run everything themselves, and that may be the best approach to succeed during the early stages of the firm's life cycle. The owners are overworked and have no free time, but that rarely changes during the early stages because their company's success depends in large part on themselves. After getting used to doing business this way, the business owner may have difficulty moving on to the growth stage, which demands the ability and willingness to attract qualified people and delegate authority to them. In the growth stage, the caliber and diversity of employees become the most important resource for continued expansion and success. In the maturity stage, the abilities of the staff remain critical and even more important than the abilities of the owners themselves.

As shown in Figure 15.2, the abilities of employees are less critical at startup and through survival, compared to during growth and maturity stages. The reason is that success in these startup and survival stages is very much dependent on the business owner who has the vision and the responsibility to execute it. It is certainly an excellent asset to have high-caliber employees, but at this stage they generally have less influence on the ultimate success of the firm, compared to their influence during the growth and maturity stages.

Once in the success stage, if the business owner decides not to grow the company, they can continue running the business themselves and the need for high-caliber employees increases slightly as the firm becomes moderately larger. If the construction firm becomes departmentalized, the importance of high-caliber employees becomes critical.

If the firm enters the growth stage, success becomes almost totally dependent on the abilities of employees. Note that this category drops off some in maturity because mature firms have fixed systems and procedures that force standardization, which reduces the impact of individuals to some extent, as long as they follow the established processes and systems. However, both in the growth and maturity stages, the abilities of the staff are even more critical than the abilities of the owners.

As discussed earlier, a mature construction company in financial difficulty reenters the survival stage. In the survival stage, having a strong independent leader is critical. A strong leader is not necessarily required in a mature company and there may not be one at the helm, but if a leader does not step forward in the event of financial difficulty, the mature company may reenter the survival stage and have difficulty managing through it.

F   *Owner's Capacity to Delegate*

During the startup and survival stages, contractors rely on themselves for success and delegate little authority to others. There may be quality people in the organization during these stages, but the prosperity of the organization lies with the owner in most cases. This becomes problematic when the owner is not willing to adapt when their company enters the growth stage. The owner's inability or unwillingness to delegate authority to others is a common cause of business failure during the growth stage. It is ironic that the attributes that are not required during the startup and survival stages are critically needed during the growth and maturity stages, where it is unlikely to succeed without delegation. The success stage can be achieved without delegation if the company plateaus at a size where the entrepreneur can still manage everything, and as long as the company does not grow significantly. However, midsize and larger companies require delegation to prosper in the success stage.

This discussion does not imply that contractors in the early stages of their company's life cycle do not delegate. It simply means that delegation is not as critical during these stages, as shown in Figure 15.2. The opposite is true if the firm grows over time into the success stage, and delegation becomes absolutely critical in the growth stage because employees will not be able to complete their functions if responsibility and authority are not delegated to them. Notice that in the maturity stage the criticality decreases somewhat because, as with quality of employees, mature firms have processes and systems that force standardization, which reduces the impact as long as employees follow the established processes and systems.

G Strategic Planning

Strategic planning becomes more and more critical as the construction company progresses through its life cycle stages, as shown in Figure 15.2. In the startup and survival stages, the entrepreneur is typically running the company and doing everything by himself/herself with limited resources. A detailed strategic plan could be viewed as restrictive, as the organization evolves based on where the market, customers, or other circumstances dictate. In these early stages, there is limited control over how, and in what direction, young construction enterprises will advance. So, one could argue that planning is of less value than in later stages. A plan could misdirect a firm to where the owner wants it to go, but where the market, customers, or other circumstances render it impractical or sometimes impossible to actually go. "Steering" a company through the obstacles and opportunities during startup and survival stages is more likely than managing it in a predetermined direction, which may partially explain the high failure rate during these early stages.

Strategic planning becomes more important for a contractor when his or her firm is in the success stage. Business planning is critical if prosperity and stability are to be maintained over time and as the organization becomes less reliant on the owner's ability. Planning is also essential in the growth stage to supervise the increasing number of people who will manage the expansion. Finding time to plan is usually an issue, but it is needed to maintain the firm's prosperity.

The mature company relies heavily on long-range strategic planning for its success and stability. Management and owners have the time to devote to planning, and it is usually a formal process, as described earlier in this book. The authors consider strategic planning a critical activity for profitable construction firms, dedicating a full chapter to the topic. In some cases, it can be said that the business plan runs the company, or at least acts as a strong guideline and control mechanism for management at every level.

A mature company pushed back into survival for any reason will have less use for its long-range strategic plan because, by definition, a company in survival has limited assurance of a future. At that point, it could be erroneous to stick to the long-range plan because it may not contain the ingredients for survival management.

H Systems and Controls

Systems and controls are how strategic plans are implemented. They ensure that what is agreed-upon in the long-range business plan is actually executed. They are, therefore, of less importance in the early stages of development such as startup and survival, because all control typically resides with the business owner. But they are increasingly significant as a construction company progresses through the later stages of success and growth. Systems and controls can become burdensome, however, if a company finds itself back in survival where fast-moving decisions must be made and implemented by a leader with limited consultation with others and without regard for existing processes.

Systems help manage the company and its employees. There are few employees in the early stages, and they typically receive instructions directly from the owner. In the growth stage, systems and controls begin to become more critical as the construction firm begins to departmentalize and expand. Systems and controls are absolutely critical until strategic planning is initiated, at which point they become less critical because strategic business planning places systems and controls into a logical and mandatory order to be executed. Strategic planning might be described as the ultimate form of systems and controls. Showing systems and controls in Figure 15.2 as becoming less critical in a mature construction company does not mean they are eliminated, but instead means that they are formalized to the extent that they are amalgamated into the strategic plan which formulates the mandatory guidelines for the whole organization to function under. So it could be said that they are still highly critical, but that they are graphically shown on the strategic planning portion of the figure, as opposed to the systems and controls chart portion of the figure, because for a mature construction company these migrate into the strategic plan.

Reverse Progression

Companies can go back into stages they have already passed through. When a mature construction company experiences growth greater than 15%, and debt increases for a year or more, it reenters the growth phase. A growth-stage company often returns to the success stage after expanding for a time. If a success, growth, or mature-stage company encounters financial difficulty where equity is plummeting rather than growing, or debt spirals out of control, it reenters survival stage. Whether management recognizes the shift or not, the shift is still real. Without understanding which stage the company is in, management may not identify which business resources are critical in that stage, and which management practices are appropriate.

It is obvious that organizations undergo changes as they develop from origination to maturity. What is less obvious is that the business resources and management practices necessary for success vary considerably between the different stages, and more importantly, that certain business resources and management practices can do great harm when applied during the wrong stage. An understanding of the stages of growth that a construction company goes through during its life cycle allows contractors and managers to identify the stage their company is in, and develop management practices appropriate for that given stage. Moreover, they can enhance their company's success potential by anticipating and identifying shifts as they occur, and then adjusting their management practices appropriately.

The Dangers of Distraction – Too Much of a Good Thing

This final concept is only prevalent in the "success" stage. Everyone reacts differently to success, and self-made prosperity can have unexpected side effects.

Entrepreneurs get totally immersed in building their business, and it becomes the central focus of their life. If success is achieved, some have difficulty if they find themselves with less to do, particularly if it is the first time in their life that they have had excess disposable income. Some have become distracted from their businesses with disastrous results. Numerous construction businesses have failed after the founder/contractor became diverted from the business, or lost interest in the business. Common diversions include golf, sailing, gambling, politics, affairs, or whatever outside interests the entrepreneur may have. In one instance, a contractor became so caught up in a trade association he spent as much as half of his time at meetings, retreats, and other organization activities with almost a cult-like attraction to the group. Another contractor, after achieving success, became involved in politics and was elected mayor of the small town he lived in, which took up an inordinate potion of his time. Predicting who these major distractions might happen to is extremely difficult, and it sometimes occurs to those least expected.

The typical subjects were self-made, hard-driving, type A personalities who spent most of their early careers building their construction businesses at the cost of personal, family, and social activities. Many had little in the way of financial advantages early in life. The distraction away from the business as the center of their life started as they accumulated considerable disposable income for the first time, with limited time to enjoy it. These contractors were typically in their 40s or older.

Hard driving at work can lead to hard playing at activities that can become addictive. Much like a midlife crisis, these activities can begin to supplant the business as the primary focus of the entrepreneur's life. Most did not notice the obsession, or they were sure that it was not affecting their business. Self-made success seems to convince people that they can do anything well, including managing their business and whatever else they would like to do.

When time and attention are directed toward new activities, it affects the concentration applied to business issues. Attention to detail diminishes, and partial or unplanned delegation too often generates unfavorable or inappropriate results. The detraction from the business comes about gradually so the impact on the business is less noticeable and often difficult to detect. When the distraction increases and becomes noticeable to others, it is seldom perceived by the contractor who generally rejects the advice of subordinates and others concerning it. In the book, *Managing the Profitable Construction Business* (Schleifer et al., 2014), this is a common element of construction business failure under the heading of "lack of managerial maturity."[1]

Prevention of this phenomenon is difficult because predicting who might be susceptible to it is complex and challenging. Convincing strong-willed entrepreneurs that they are at risk from this is extremely difficult, and intervention once the distraction occurs is usually aggressively resisted. Many small and midsize contractors have little accountability to anyone, so the institution of a Board of Advisors or Board of Directors is an excellent prevention against diminishing interest of the founder or leader. As described elsewhere in this book, an independent board is in a position to observe changes sooner, and if trusted and relied upon, may be in a position to impact the outcome.

Review Questions. Check All That Apply.

1 The failure rate for startup construction companies:
   a Varies over time
   b Is 30% to 50%
   c Is 50% to 70%
   d None of the above

2 What do the authors list as a key ingredient during startup stage?
   a Optimism
   b Perseverance
   c Endurance
   d All of the above

3 Growth over ___ can shift a company from one stage to another:
   a 15%
   b 25%
   c 35%
   d 45%

4 In which stage are business connections most important?
   a Startup
   b Survival
   c Success
   d Growth

5 In what stage(s) are the dangers of distraction prevalent?
   a Survival
   b Success
   c Growth
   d Maturity

Critical Thinking and Discussion Questions

1 Explain why a company cannot choose or decide what stage it is in.
2 Discuss the difference between success and growth stages.
3 Why is the owner's ability critical in startup and less important in maturity stage?
4 Explain how a company can ship a stage in reverse progression.
5 What is your opinion of the section on dangers of distraction?

Note

1 For a more in-depth presentation of each element refer to the book *Managing the Profitable Construction Business*, by Thomas C Schleifer, PhD, published by Wiley, 2014.

16 Succession Planning

A closely held construction company can be defined, at any given point in its life cycle, as the sum and substance of its experience. Almost all of that sum and substance has been created by, and resides in, people, particularly and proportionately in the firm's senior managers. Sum and substance includes talents, abilities, and attributes of a manager. When key people leave a closely held construction enterprise expectedly, such as in planned retirement, or unexpectedly, such as in illness, some of the "essence" of the organization is lost.

Part of the sum and substance becomes institutionalized, including general attitudes toward customers, work ethics, values, and so on, because the key people have woven them into the "fabric" of the organization, by mentoring, training, and leading by example. A portion of the sum and substance can be transferred from one person to another with deliberate and diligent effort; these include customer relationships, union contacts, production and process knowledge, and so on (Figure 16.1). However, a considerable portion of the sum and substance cannot be institutionalized, is not easily transferable, and, in spite of the leaders' best efforts, is forever lost to the organization. When a key person leaves an organization, particularly a founder, the firm may lose insights into business or project risks, the development of innovative processes, methods and actions born out of years of company-specific experience, masterful hiring selections, and solutions to personnel issues, resulting from an ability to read between the lines or "hear" what is not being said, using an innate talent honed over decades.

The Basics of Succession Planning

To salvage as much of the sum and substance as possible, it is best to have around three years to enact the process for middle managers, and around four to five years for senior managers who will be leaving an organization. For key senior management departures, five years of preparation is appropriate. This timing is not always possible, but when it is, it can significantly improve the outcomes for the organization. If a key person leaves without notice, the same principles apply, but the risk is much higher. When there is uncertainty about when a person may leave, such as an intention to retire in a non-specific length of time, or uncertainty

DOI: 10.1201/9781003229599-20

Figure 16.1 A portion of the sum and substance can be transferred from one person to another with deliberate and diligent effort.

about health issues, it is prudent to plan to the shortest time. If the person ends up staying longer, that's even better.

The timing outlined above provides the luxury of the person leaving being able to train, teach, and coach their replacement into the new position. During the final year or six months of this period, it is recommended that the replacement be promoted into the position with full responsibility. There is a considerable advantage of having the retiring person still there to mentor, assist, and encourage. If any or all of this sounds like more effort than necessary, or perhaps too expensive, consider the cost of getting it wrong and having to do it over again. That may not even be possible. It should also be understood that the selection of an individual to be advanced or promoted has the potential to cause emotional reactions from others, particularly some who think they should have been selected instead, which can lead to early departures of disgruntled employees. This potential disruption of the management team during the succession process can further complicate the issue.

The Risks Associated with Succession

The construction business failure rate at succession is in excess of 50%, with one study estimating a 65% failure rate at that juncture. Succession might be thought of as a "silver bullet" for the successful construction firm. At a minimum, the succession process is to be respected as a high-risk event; some would even call it a "dangerous" event. Succession must be diligently planned for, well in advance of a key person leaving the organization, and must be well executed, ideally while the key person is still with the company.

Risk of Sudden Departure

If departure is sudden, the firm is immediately at extreme risk. A good defense strategy when faced with a sudden departure of a key person is to reduce sales volume immediately by about 15%, and preferably 25%. The purpose is to allow greater concentration on less work into the future until the impact of the sudden absence of the key person is able to be evaluated. No matter what actions were taken to replace the person, the results of the "new" organization will not be known for at least a year of working together. It takes at least that long to measure performance against prior years.

A serious reduction in volume reduces the strain on the remaining managers and company resources as the succession process unfolds and allows the new team to be tested under less pressure. This is critical, as experience shows there is less than a 50% chance that the succession will be successful. If it does not work, that may not be visible for a year or more, at which time the succession process will have to be repeated. There is clearly less risk if the firm is not fully loaded with work. This type of circumstance is one of the most serious crises the firm will ever face, and a good first reaction must be to downsize, analogous to saving a sinking ship by lightening the cargo load.

Succession in Family Businesses

Succession in family businesses has its own unique risks because the successor is often chosen by birth, and not necessarily qualifications. This is not a critique, nor a suggestion that family members are automatically not qualified. Owners have the right to select whomever they like for senior positions. What tends to happen is, with the successor a foregone conclusion, few firms utilize a well thought-out succession plan, diligently managed, supervised, and measured. The family member successor incurs a huge disservice when the succession planning and implementation are skipped. The family member is put at a disadvantage without the value of a group of senior managers evaluating, developing, and managing the preparation for the new position. Compare that to competitors in non-family-owned businesses, where a non-family member would normally be the recipient of this intentional development and coaching process. The disadvantage is compounded when some are put into a senior position at a very young age, especially positions for which a non-family member would never be considered because of lack of experience. The ability to manage a construction enterprise successfully is not inherited.

The retirement of a founder of a small or midsize construction business, particularly a first-generation founder, has exposures that may not be anticipated. As described in the previous chapter on stages of growth, founders of most construction companies are typically self-made contractors that lead and direct everything and have not had to delegate much authority during the startup, survival, or success stages of their businesses. As a result, they may not delegate enough authority to their family successor to learn enough about running the business and being successful doing so. With the prosperity of the founder, the successor has usually spent years in higher education before joining the firm, and then often as a trainee in the office. When this is the case, there is usually minimal field experience, a lot of which was gained by watching, not by doing. As explained in the stages of growth chapter, companies in each stage, other than maturity, require a hands-on leader/entrepreneur to be successful, which may not be within the training and experience of the successor, exposing the firm to serious succession risk.

Contractors who have achieved the success stage have another risk exposure that might be described as "too much of a good thing." Prosperity in the success stage, combined with a reduction in the demands on the founder's time, has led to distractions from the business. Distractions can create a leadership vacuum, meaning the founder is partially pulling away from the business without a successor in place. This risk was addressed in more detail in the previous chapter of this book.

The Succession Planning Process

The succession plan is developed by a small group of top managers, and may include the candidate who will replace the executive leaving the firm if that candidate has already been selected. The plan should outline the sum and substance that may be lost as a result of the executive leaving, how that will be institutionalized or transferred, and to whom. It should also clearly define what portion of the sum and substance will potentially be lost to the organization, and how that may be compensated for in the future. It should lay out what specific actions will be taken during the duration of the plan, by whom, by when, and in what sequence. The plan should contain as much detail as the planning group believes is necessary. This chapter provides a sample succession plan, which is only intended as a guide to what might be included in a plan. The content of the plan will vary considerably from one firm to another. A senior member of the team that developed the succession plan should be selected to manage the plan through its duration and mentor the candidate one-on-one.

The steps necessary to minimize the loss of sum and substance to the organization are referred to as a "succession plan." The succession planning process includes:

1   Discovering, evaluating, and delineating exactly what sum and substance is embedded and engrained in a key person well before they leave the organization.
2   Determining which of these elements can be institutionalized, and outlining steps to be taken to ensure that they will be instilled and firmly engrained into the "fabric" of the organization.
3   Identify the sum and substance that can be transferred to another person, determine who they should be transferred to, and devise a method for

systematically transferring them over time, including by whom and by when. This process usually takes three to five years, depending on the seniority and role of the person leaving.

4   Establish which sum and substance are not transferable and will be lost to the organization. Minimize the resulting impact by recognizing which missing talents, abilities, or attributes will not be applied to some future business decisions. Realize that in each instance, management will need to evaluate the impact that missing ingredients may have on the risks associated with future decisions, and attempt to compensate for the increased risk by introducing additional people and collaborations before making the decision.

5   Cultivate similar talents, abilities, and attributes, to the extent possible and practical, in the successor of the key person who is leaving, through a carefully developed and diligently pursued multiyear professional development plan (PDP), as described in the remainder of this chapter.

A Sample Succession Plan

The following succession plan (Figure 16.2) is intended as a guide because plans vary considerably from one firm to another. The format and content of the sample plan should be altered and adjusted to suit the needs of the planning group and the candidate. It should contain as much detail as the group deems necessary and appropriate. The plan should outline the sum and substance that may be lost to the organization as a result of the person leaving. It should define the steps that will be taken to institutionalize what can be institutionalized, transfer what can be transferred, and recognize those talents that will be lost. Within the plan, preparing the replacement is as important as determining which sum and substance will be lost, how the organization will be impacted in the future, and how the entire team will compensate for the lost talent, style, contacts, and so on.

One of the senior managers involved in the preparation of the succession plan should be selected to manage the plan through its entire multiyear duration and to mentor the candidate one-on-one. If qualified, appropriate, and willing, the key person planning to leave the organization could be the one managing the plan and mentoring the candidate. This activity requires a non-aggressive personality and the ability to establish a comfortable working relationship with the candidate. The team's selection of the individual managing the succession plan must be made with care, as it is critical to the plan's success.

Managing the Professional Development Plan of a Leader-in-the-Making

The PDP is intended to provide information, increase knowledge, and open lines of communication to prepare the candidate for the replacement position. This may include mentoring, coaching, continuing education, introductions to counterparts for sources of information and advice, books, articles, magazine and journal subscriptions, and anything else that can advance the candidate's preparation for their new role in the organization.

PDPs, skills inventories, questionnaires, and communications concerning these should be considered confidential as they contain personal and professional information about an employee. The amount of personal time and energy the candidate devotes to their professional development activities will vary and depend on the individual candidate, because employers should not attempt to control an employee's personal time. Their progress is a measure of their ambition as well as other personal factors.

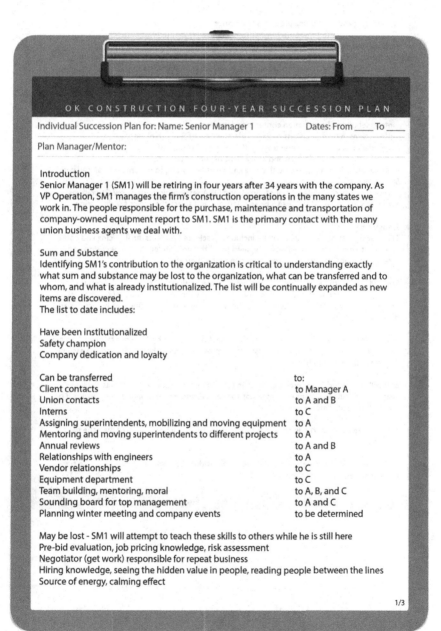

OK CONSTRUCTION FOUR-YEAR SUCCESSION PLAN

Individual Succession Plan for: Name: Senior Manager 1   Dates: From _____ To _____

Plan Manager/Mentor:

Introduction
Senior Manager 1 (SM1) will be retiring in four years after 34 years with the company. As VP Operation, SM1 manages the firm's construction operations in the many states we work in. The people responsible for the purchase, maintenance and transportation of company-owned equipment report to SM1. SM1 is the primary contact with the many union business agents we deal with.

Sum and Substance
Identifying SM1's contribution to the organization is critical to understanding exactly what sum and substance may be lost to the organization, what can be transferred and to whom, and what is already institutionalized. The list will be continually expanded as new items are discovered.
The list to date includes:

Have been institutionalized
Safety champion
Company dedication and loyalty

Can be transferred	to:
Client contacts	to Manager A
Union contacts	to A and B
Interns	to C
Assigning superintendents, mobilizing and moving equipment	to A
Mentoring and moving superintendents to different projects	to A
Annual reviews	to A and B
Relationships with engineers	to A
Vendor relationships	to C
Equipment department	to C
Team building, mentoring, moral	to A, B, and C
Sounding board for top management	to A and C
Planning winter meeting and company events	to be determined

May be lost - SM1 will attempt to teach these skills to others while he is still here
Pre-bid evaluation, job pricing knowledge, risk assessment
Negotiator (get work) responsible for repeat business
Hiring knowledge, seeing the hidden value in people, reading people between the lines
Source of energy, calming effect

1/3

Company psychologist, cheerleader, second chances
Exploring new frontiers
Experience with all the types of work we do
Anticipating and identifying problems in advance

Training and Transfer of Knowledge to Successors
The three Project Managers A, B, and C will be promoted to Operations Managers immediately so that SMI can transfer portions of his project responsibilities to them and concentrate more time on training, mentoring and transferring proprietary knowledge and his business contacts to them as well as to estimators and others. The transfer process will include:
- regular ride-alongs
- introduction to clients, union contacts and vendors
- interactions in field decisions to gain understanding of SMI's thinking
- attendance at association meetings, golf outings, vendor events, etc.
- participation in the selection and movement of superintendents
- participation in company-wide thinking and decision making to become acquainted with SMI's thinking
- twice a month scheduled 2-hour training and Q&A sessions with each new Operations Manager. Operations Managers to bring questions and select the topics for discussion.

Transfer Process
The new Operations Managers will be included in selected projects' final pricing and coordination with estimating, to become prepared to fill that role under SMI's supervision and tutelage in year two and three, then independently thereafter.

SMI will include the appropriate operations manager in all interaction with candidates for employment, interviews and hiring decisions, and specifically teach and mentor the managers on how he handles those issues, what he is thinking, and what influences his actions and decisions.

SMI will include the three managers, to the extent practical, in his "Senior Statesman" role in the organization so that they can assist with that over the next three years and can assume the role after that timeframe.

SMI will also include the three managers in his "big picture" ideas and share his thinking, insights and "gut" reactions in an attempt to preserve some of the "intangibles" that will be lost to the organization.

A new more delineated equipment assignment protocol will be developed over the next two years.

The assignment of projects and customers to individual Operations Managers will be transferred to the Senior VP Estimating.

2/3

Professional Development of Successors
Skills will need to be developed and measured in the broad areas of technical skills, leadership, business/finance, and marketing. The minimum skill sets under each heading are:

Technical: An understanding of equipment utilization/management, concrete forming systems, characteristics of concrete and mix design, planning and scheduling, productivity measurement principles, estimating takeoff/pricing, field safety, OSHA regulations. An exposure to soil mechanics, strength of materials, and basic physics.

Leadership: An understanding of leadership styles, leadership ideologies, principles of motivation, differences between management and leadership, use of authority and power, impacts of change, maintenance of trust coalitions, significance of vision.

Business/finance: An understanding of basic accounting, bank credit granting principles, how surety works, importance of cash flow, business planning methods, basic contract law, contract administration, construction cost accounting, alternative dispute settlement techniques. Exposure to basic statistics, basic risk analysis, insurance principles, business ethics, computer technology, General Accepted Accounting Principles (GAAP).

Marketing: An understanding of basic marketing principles, public relations, the persuasiveness of selling, marketing intangibles and the importance of measuring the competition, market analysis principles, types and purpose of advertising, concepts of consumer behavior.

Assumptions
The candidates have mastery of written and oral communication skills, are computer capable, have good people skills and have been educated in basic accounting methods and principles, micro-economics and macro-economics, business law, basic psychology, report writing, business research methods, management theory, personnel management.

Learning Methods
Skills can be learned and/or enhanced through on-the-job training, in coursework or self-taught, primarily through reading. Some skills are learned more efficiently or effectively through one method rather than another. It is believed that at least some coursework (evenings) will be necessary, but attempts will be made to minimize that method. Alternative methods will be suggested where possible. Experience, by definition, can only be gained by doing; but carefully designed field and office experiences that are vigorously pursued is a learning alternative. A Professional Development Program (PDP) will be developed for, and in consultation with, each candidate to accelerate their learning. The first step will be a self-evaluation of current skill sets, proficiencies, and interests. The PDPs will be customized to needs perceived by the candidate or SMI, and candidate preferences and interests.

3/3

Figure 16.2 Sample of four-year succession plan.

The more senior the person or the greater the amount of sum and substance presumed to be lost, the greater the effort and resources to be applied to the development of the succession plan. It is appropriate to alter the plan to address the needs of the candidate as he or she progresses, and make changes determined by the mentor and the candidate or as a result of the mentor's interpretation of the periodic evaluation questionnaires.

Skills Inventory

To initiate the professional development process in preparation for developing a PDP, it is helpful to have a skills inventory of the candidate. While the skills inventory is not an objective measurement, in part because it is self-administered, it is helpful to understand what the candidate believes their skill level is in various areas. The sample skills inventory shown in Figure 16.3 is for the candidate in the sample PDP presented later in this chapter (Figure 16.4), so the questions parallel the skill sets management believes are significant in the position under consideration. The subjects would of course be customized to fit the circumstance.

Measuring Succession Plan Progress

Several months into the succession plan it is appropriate to measure progress. One method of self-evaluation and measurement of progress by management is to use a short questionnaire, such as the template provided toward the end of this chapter (Figure 16.5). Those intended to advance in their responsibility and those in training should fill out the questionnaire. Note the questions parallel the plan elements. The template questionnaire is embedded in a sample memo to the candidate who will be evaluated. It should be modified to fit the circumstances and customized to the candidate and the position being explored.

Succession Planning and Risk Management

Succession planning is critical for any key person planning to retire or to leave the firm (Figure 16.6). It is also prudent to develop abbreviated succession plans for all key managers in a firm for the possibility that they may leave unexpectedly for health or other reasons. While succession planning is most important at the executive level, the concept applies more broadly in the company. This planning causes management to take into consideration the sum and substance each key person possesses and evaluate if any portion of their job description or responsibilities can or should be institutionalized or shared with others for the eventuality of a sudden departure. The planning activity causes speculation about who would replace that person and if a qualified and suitable person is even employed within the firm. This does not mean you would necessarily begin a search for a replacement, but may cause you to instigate a better distribution of the sum and substance more widely within the organization to the extent possible. This type of succession planning is an

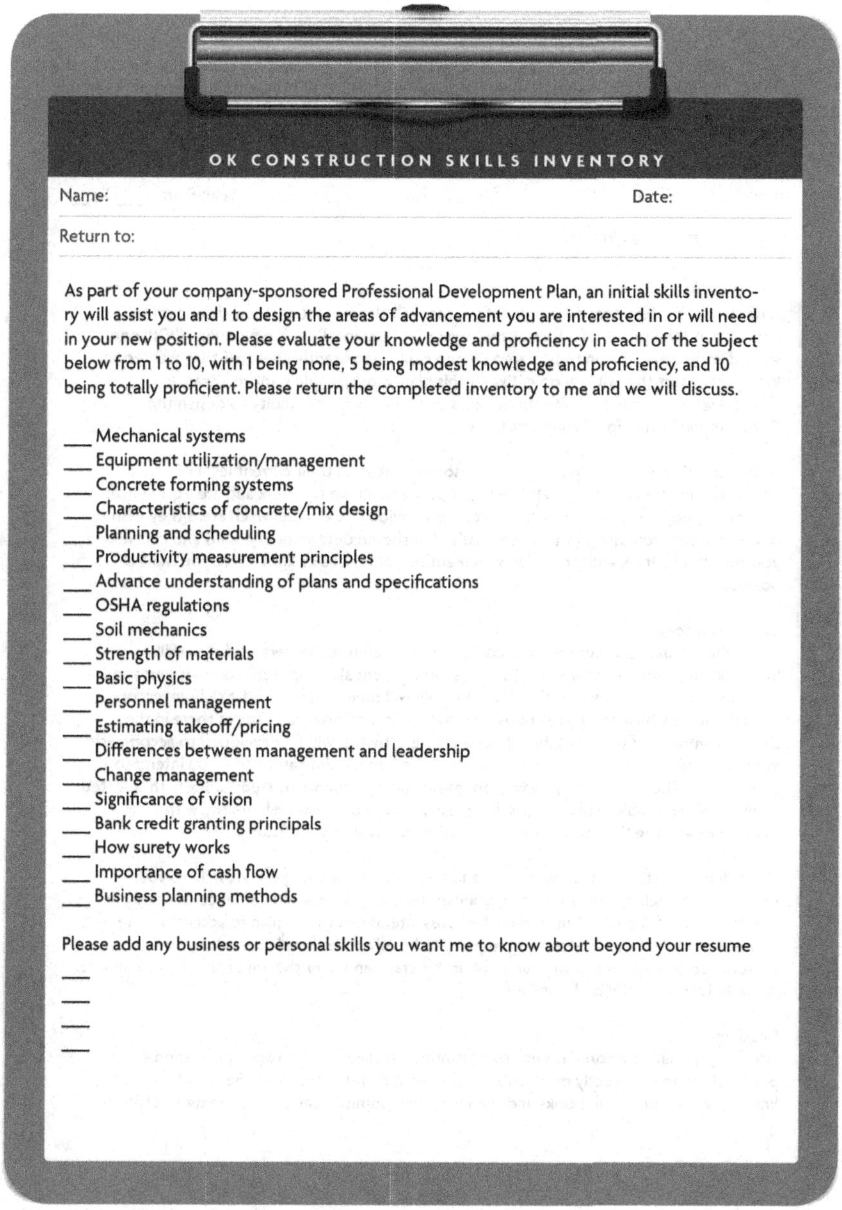

OK CONSTRUCTION SKILLS INVENTORY

Name: _____   Date: _____

Return to: _____

As part of your company-sponsored Professional Development Plan, an initial skills inventory will assist you and I to design the areas of advancement you are interested in or will need in your new position. Please evaluate your knowledge and proficiency in each of the subject below from 1 to 10, with 1 being none, 5 being modest knowledge and proficiency, and 10 being totally proficient. Please return the completed inventory to me and we will discuss.

___ Mechanical systems
___ Equipment utilization/management
___ Concrete forming systems
___ Characteristics of concrete/mix design
___ Planning and scheduling
___ Productivity measurement principles
___ Advance understanding of plans and specifications
___ OSHA regulations
___ Soil mechanics
___ Strength of materials
___ Basic physics
___ Personnel management
___ Estimating takeoff/pricing
___ Differences between management and leadership
___ Change management
___ Significance of vision
___ Bank credit granting principals
___ How surety works
___ Importance of cash flow
___ Business planning methods

Please add any business or personal skills you want me to know about beyond your resume

___
___
___
___

Figure 16.3 OK Construction's skill inventory.

CONFIDENTIAL
FOUR-YEAR PROFESSIONAL DEVELOPMENT PLAN (PDP)

Name of Candidate: _____   Year: From ____ To ____

Name of Plan Manager/Mentor: _____

Introduction

A multi-year professional development process will also impact personal development. To comprehend the power of a formal, multi-year Professional Development Plan (PDP) one should understand that in four years a person could significantly change if they wanted to. While this is not the primary objective of this plan, most people want to advance themselves over time, and with this process you have the opportunity to design the Construction Professional you want to be in the future.

A formal PDP is intended to advance a person well beyond their current level of understanding through accelerated learning over and above the work and life experiences that most people gain each day. It is up to the individual to decide which areas they wish to advance, both professionally and personally. You should determine the skills and knowledge you wish to enhance, and/or decide which entirely new skills or knowledge you want to possess.

Skills Inventory

To establish a baseline, current skills and interests need to be understood. The skills inventory you completed was used in evaluating current skills, aptitudes and interests and provided guidance in the plan development. We will meet to go over the skills inventory and to discuss which areas you choose to advance in, and how you can get there in the shortest amount of time. A schedule of learning activities will be developed to accomplish your goals, after which you will need to decide how much time and effort you intend to devote to it. The plan will rely heavily on reading and independent study along with selected seminars and possibly short courses. I suggest you set your goals in harmony with your aspirations and the time commitment you wish to invest in your future.

You will be investing considerable time and energy to accelerate your growth, so your thoughts and feelings are critical to the initial decisions and should influence the content and "texture" of the plan. You and I will discuss alterations to the plan to accommodate your requirements now and as you progress. To be effective this plan needs to be customized to your needs, so your input at this stage and over the life of the effort will have acute influence on the path forward.

Reading

Life-long learning is a commitment to continuous advancement in one's professional and personal life and is directly proportional to a person's determination. The world body of knowledge is captured in books and the ultimate recommendation is to create an attitude

1/4

and/or change behaviors so that reading becomes a permanent habit. Many people avoid reading non-fiction text material because they consider it boring and hard to get through. However, when something you learned from a book solves or avoids a problem, your job becomes easier, and when the reading effort pays off it seems a lot less difficult. Serious managers read as many as six or eight non-fiction books a year. We recommend a minimum of three such books per year after this plan is completed as part of continuing life-long professional growth.

To kick start the learning process and to maximize achievement from initial efforts I suggest we start with some books that address subjects you will be advancing in.
The initial list of recommended books is below.

Personal Development
Body Language; Julius Fast – This is old and basic, but I like it. If you find a later text on the subject it will probably be as good.
Unlimited Power; Anthony Robbins – Good for self-motivation
Leadership
Leadership Secrets Of Attila The Hun; Wes Roberts – a classic
Seven Habits of Highly Effective People; Steven Covey – a classic
Management
The One minute Manager; Kenneth Blanchard – A classic easy fast read.
Business Development
Raving Fans; Kenneth Blanchard – Another easy to read classic.

Personal Resource Material
I strongly recommend that you begin a personal reference library and get into the habit of reading with a highlighter in hand and highlight phrases or sentences that interest you or facts you may want to remember or refer to later. Make notes directly on the pages. In the unlikely event you come across words you are unfamiliar with, look them up and this will also be a vocabulary-building exercise. We often think we will remember what we read if it is informative or interesting, but it escapes us over time. Highlighted and marked up books that you have read are a tremendous resource in problem solving that you can refer to throughout your life to recapture information you can't quite remember, but know you have seen somewhere. If you prefer, feel free to use a tablet to electronically save your notes, highlights, bookmarks, and so on.

Trade Publications
Another valuable source of information are trade magazines and journals. I suggest Engineering News Record at a minimum, the American Concrete Institute trade association magazines, and any others of your own selection. At least one business magazine such as Forbes, Business Week, or any others. They do not have to be read carefully but should be scanned for articles and subjects that you find of interest. Even then, the articles need not be carefully read if interest or time does not permit. Tear them out and file under general

headings like equipment, tools, products, management, leadership, etc. Use a system of file headings that makes sense to you and this will become an invaluable personal reference source. You can then find reference material when working on a subject or problem and recover information you already have, providing data and insights into the topic at hand. Many use their personal reference source regularly when writing reports or solving problems.

These are good habits to get into even if you may not see the value up front. They are proven techniques that work well with earnest reading habits, and even better after you have collected some information in your files and on your bookshelves. When you start the last book on the list, let me know and I'll send more titles.

Self-Study
Many skills, particularly communication and negotiating skills can be advanced in self-study. After you read the body language book, put into practice what you learn. Become more conscious of studying people more thoroughly through their actions as well as their words when you interact. This is an excellent skill to perfect. It is valuable in managing, marketing, negotiating and presentations. After you try this for a while, if the subject has more than a passing interest to you, we will include another book on it in the next selections and we will move on to the study of group dynamics.

We recommend you become modestly active in one or more trade associations because just being with other construction professionals is a learning experience. The firm is a member of a number of association; mingling at local chapter activities has the added advantage of meeting sources of information and potential clients. The company is a member of a number of other trade associations which you may want to glance at the mailings, newsletter, etc. These types of activities and relationship can bring you into the "fabric" of the regional construction community. Because of time constraints you may need to start slow, but we should include trade associations in the plan.

We should talk about effective subcontract management techniques, change order management, customer service and marketing, and estimating issues, to advance your knowledge in those areas. I suggest we develop a schedule of the variety of activities you would like to become familiar with, such as the customer service and marketing meetings you should attend as a learning activity. These should be listed and scheduled, including a determination of who you will learn from. The options are to address one area at a time for a given period of time, say three months, or all at once as opportunities arise. This should be enough to get us started and we can expand the list as we go.

Learning about these subjects will also include selected readings, mostly copies of articles I will send you. You will need to understand the company's subcontractor management philosophy and all field-related policies. You should study the written corporate policies because one of the attributes of the well-managed company is consistency which is

3/4

fostered by written corporate policies along with better quality control and risk management.

Seminars And Conferences
These need to be selected as they become available. I recommend you attend two or more per year. Most are announced months in advance, so it is an ongoing selection process. You and I can look for available programs.

Short Courses
There are a number of one-week and two-week powerful courses available that may form a part of this plan, but not the first year and not until definitive progress is made. We may want to identify several courses and plan for you to find time for them in years three and four. They are excellent and very appropriate but time consuming so we will address them carefully and not necessarily initially.

Follow Up
I will follow-up with you regularly. It will be helpful for both of us if you periodically present a written or oral report on your progress, perhaps every three months. It will help you do a self-assessment and we can discuss your development and needs so that I can assist in your efforts.

A Continuing Process
A multi-year Professional Development Plan is not set out in its entirety because the pace of professional development is set by the participant. It is also dependent on the availability of learning opportunities, and the bodies of knowledge we will be drawing on can change over time. The plan will be updated periodically in consultation with you. I will be available for assistance and consultation, but most of the interaction should be prompted by you as you see the need. Your plan and progress will be confidential between you and I, however you are certainly at liberty to discuss it more widely if you chose.

4/4

Figure 16.4 Sample of professional development plan for a leader-in-the-making.

OK CONSTRUCTION SUCCESSION PLAN
PERIODIC EVALUATION

To: (candidate)   Date:

From: (mentor)

The company is investing in assisting you to accelerate your growth to assume the responsibilities of (name of person retiring) SM1_____ so please take the time to think these questions through. Your thoughts and feelings in each area are critical to the evaluation process and will form the basis for you and I to discuss your company-sponsored Professional Development Plan progress (unless this is the initial evaluation). After this we will discuss progress and any alterations to the plan to accommodate your requirements, or if either of us thinks appropriate changes or improvements to your professional development can be made. This plan is customized to your needs so evaluation of progress at this stage will influence where we go from here.

Please answer the following questions on a scale of 1 to 10, with 1 being little progress, 5 being some progress, and 10 being excellent progress. Feel free to add comments that you feel apply.

Please return the completed form by _____[DATE HERE]_____

| 1 | 2 | 3 | 4 | 5 | 6 | 7 | 8 | 9 | 10 |

Little progress   **Some progress**   **Excellent progress**

___ Progress in being introduced to client contacts   Comments:

___ Your progress in establishing relationships with new clients
Comments:

___ Your progress in managing and improving existing client relationships
Comments:

___ Progress in being introduced to engineers
Comments:

___ Your progress in establishing and improving relationships with engineers Comments:

___ Progress in being introduced to vendors
Comments:

___ Your progress in establishing and improving relationships with vendors
Comments:

___ Progress in attendance at association meetings, golf outings, vendor events, etc.
Comments:

___ Progress in being introduced to union contacts
Comments:

___ Your progress in establishing and improving relationships with union contacts
Comments:

___ Your progress in understanding how to assign superintendents—what to consider
Comments:

1/2

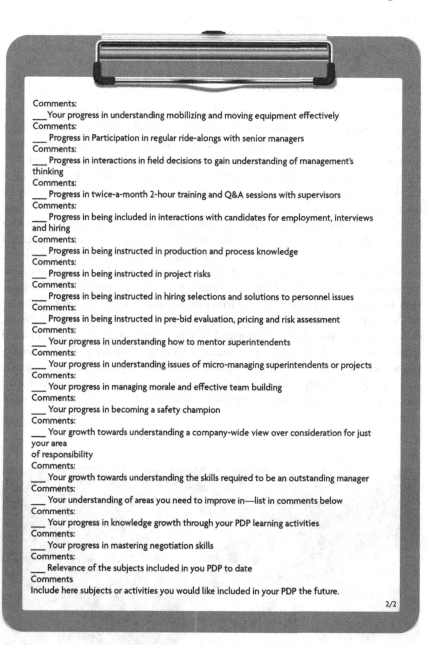

Comments:
___Your progress in understanding mobilizing and moving equipment effectively
Comments:
___ Progress in Participation in regular ride-alongs with senior managers
Comments:
___ Progress in interactions in field decisions to gain understanding of management's thinking
Comments:
___ Progress in twice-a-month 2-hour training and Q&A sessions with supervisors
Comments:
___ Progress in being included in interactions with candidates for employment, interviews and hiring
Comments:
___ Progress in being instructed in production and process knowledge
Comments:
___ Progress in being instructed in project risks
Comments:
___ Progress in being instructed in hiring selections and solutions to personnel issues
Comments:
___ Progress in being instructed in pre-bid evaluation, pricing and risk assessment
Comments:
___ Your progress in understanding how to mentor superintendents
Comments:
___ Your progress in understanding issues of micro-managing superintendents or projects
Comments:
___ Your progress in managing morale and effective team building
Comments:
___ Your progress in becoming a safety champion
Comments:
___ Your growth towards understanding a company-wide view over consideration for just your area
of responsibility
Comments:
___ Your growth towards understanding the skills required to be an outstanding manager
Comments:
___ Your understanding of areas you need to improve in—list in comments below
Comments:
___ Your progress in knowledge growth through your PDP learning activities
Comments:
___ Your progress in mastering negotiation skills
Comments:
___ Relevance of the subjects included in you PDP to date
Comments
Include here subjects or activities you would like included in your PDP the future.

2/2

Figure 16.5 Sample memo to candidate, with evaluation questionnaire.

excellent exercise in risk management as it exposes typically unexplored risk of the sudden departure of a key person. The risk may be minor, but it's better for it to be a known risk than an unknown one.

Follow-Up After Succession

When the succession is accomplished and the key person leaves the firm, it is helpful to their replacement if they can be accessible periodically for about a year. Depending on the retiring manager's health, availability, and willingness, it is of value to make arrangements for them to be reachable by phone, virtually, or in person to answer questions or offer advice to their replacement if asked for. This may require compensation but has great value if it can be organized.

The small and midsize construction enterprise is a complex, high-risk business demanding a unique set of skills found in successful contractors. Those lacking the skills, drive, and willingness to work continuously are often weeded out, while others sacrifice everything for the business with no guarantee of success. The high failure rate following succession in construction businesses emphasizes the critical importance of the founder or leader to the success of a construction enterprise. When failure follows succession, it is often because the successor did not have one or more of the critical ingredients of skill, drive, or commitment that the founder had. Implementing a succession plan, including a professional development program for the successor(s), can help reduce and manage some of the risks associated with succession in construction enterprises, increasing the probability of a positive outcome.

Figure 16.6 Succession planning is critical for any key person planning to retire or to leave the firm.

Review Questions. Check All That Apply.

1 The sum and substance the authors describe include:
   a Talents
   b Abilities
   c Attributes
   d All of the above

2 A good defense strategy when faced with a sudden departure of a key person is:
   a Move someone temporarily into the position
   b Use a temporary worker
   c Reduce sales volume
   d Have a senior person step in

3 What do the authors say the succession plan should outline?
   a The sum and substance that may be lost
   b How the sum and substance should be institutionalized
   c How the sum and substance should be transferred
   d All of the above

4 Professional development programs, skill inventories, questionnaires, and communications concerning individual are:
   a The building blocks of success
   b Are able to define a person
   c Considered confidential
   d Required of each manager

5 When failure follows succession, it is often because the successor did not have the necessary:
   a Skill
   b Drive
   c Commitment
   d All of the above

Critical Thinking and Discussion Questions

1 Describe and clarify what the "sum and substance" means to you.
2 Discuss the issues you consider most important in succession of a family business.
3 Explain the need for and advantages of a succession plan.
4 Expand on the message in the "follow-up after succession" section.
5 Explain the message in the section of managing the professional development plan.

17 Professional Development

Decades ago, the construction industry did not formally practice continuing education. With time, such activities were considered fine additions, but not necessary. Today, continuing education is a necessary for a contractor's success (Figure 17.1).

In our experience, construction industry skills (including management skills) have a half-life of five years or less, which means that about half of the skills workers use to do their job will be obsolete to that job in five years, and most skills will need to be replaced with new skills every ten to fifteen years. The construction industry lagged behind the continuing education and training thrust of the 90s and was late in recognizing that maintaining the skill levels necessary for executives, supervisors, managers, and the labor force to do their jobs is in large part the responsibility of the company, not the employee. Inefficiency and unimpressive productivity in the construction industry were regularly tolerated behind the excuses that every project is different, there is limited control of the work environment, and the labor force is not consistent from project to project.

The Need for Continuing Education in Construction

Back when construction was still considered a custom product, the average contractor enjoyed sufficient margins to profit in spite of built-in inefficiencies. As construction moved closer to becoming a commodity in many sectors of the industry, efficiency and productivity became critical to profitability. Larger companies started to embrace training and education as a necessary cost of doing business. They have the advantage of size, and economies of scale help cover the cost of the needed resources and upfront expenses. Small and midsize organizations can accomplish this through their trade associations, but time is short. Larger companies are moving quickly, and small and midsize companies are not always known for collaboration and joint action.

Many contractors consider continuing education and training expensive. However, the term *expensive* is relative, and education and training typically provide a superb return on investment. The rapidly changing business environment requires increasing sophistication and efficiency to remain competitive, and many construction enterprises are playing catch-up to overcome skill

DOI: 10.1201/9781003229599-21

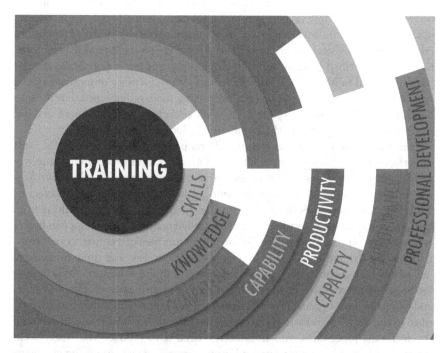

Figure 17.1 Continuing education is necessary for a contractor's success.

deficiencies. Some contractors are starting to view education and training as a wise investment rather than an expense.

Company-Sponsored Education and Training

One reason company-supported continuing education and training may seem expensive is that in the past that was not a regular cost of doing business. Contrast this to banks, for example, who train people to be bank-tellers because they do not expect people to come out of school knowing how to be tellers. Banks continue training on a regular basis as their systems, procedures, policies, and regulations change, and they continue to provide their employees with the education and information to advance within the firm. The construction industry has evolved, becoming more complex to the point that the successful contractor of the future has no choice but to embrace continuing education and training as a cost of doing business if they expect to succeed in construction. The organization that delays this investment or disputes this reality will remain static as competitors advance, resulting in shifts in an already competitive marketplace.

The timing of this need is critical as profit margins decrease while industry business risks remain constant. This elevates the demand for efficiency and productivity from essential to survival-level. Our industry has often overestimated on-the-job training and years of service, overvaluing some people with 25 years of experience who in fact have five years of experience and 20 years of repetition. An organization of professionals who do not regularly renew their skills and learn new ones is destined to replicate itself indefinitely, missing opportunities for improvement and falling behind the efficiency and productivity of more advanced competitors. Large companies are not necessarily more enlightened, but seem to better recognize the necessity for skills development and enhancement and are generally willing to invest in their employees' education and training.

Even modest education and training efforts provide huge returns in the form of performance enhancement, employee motivation, and productivity improvements. Most companies invest in executive and management education first, which does not offer direct measurements of field productivity enhancements. Therefore, few companies have a baseline for quantifying changes resulting from training costs, but all report recognizable improvement. Our experience indicates education or training expenses applied in almost any area of a construction organization are recovered rapidly, and in a number of cases, two and three times over.

The time for debating the relative value of continuing education and training has been overtaken by the rapid rate of change in the business atmosphere. Owners and designers have become more sophisticated, and construction has been demystified to the degree that most of the parties to the construction process understand the methods, procedures, and information needed to produce the work. Well-informed designers, owners, users, and inspectors provide an increasing challenge to construction professionals who have little choice other than to improve their skills, or fall behind. On-the-job training is not enough. Improvement and replacement of skill sets require serious study and a commitment to life-long learning.

The construction industry is at a crossroads. We can continue to debate, or we can provide necessary education and training to keep up with the business environment. The question, "how much will it cost?" will inevitably be asked. The answer is: "as much as an organization can afford." There is a huge void, and the first to improve will shift competitive balance and could pick up a lead that will be hard for others to overcome. The reason is clear; education and training pay immediate dividends in the form of increased profits. Contractors who invest will maintain or increase their lead by reinvesting profits resulting from education and training into additional education and training. There is tremendous opportunity in continuing education, accelerated by the increase in the rate of change the industry is experiencing, which is magnified by the neglect to date.

For many years, construction has been described as an industry of brawn and determination, with a reputation that stems from building custom products, with rudimentary equipment and limited technical support. Our industry resisted change, was slow to embrace technology, and staked its reputation on a can-do attitude. New technology is upon us, driven more by the skilled labor shortage than a willingness to change, while innovative systems and processes have simplified the production and organization of the work.

In many sectors, construction is (or may eventually become) a commodity. The production of commodities demands efficiency. Builders will either be numerous, well-educated, and well-trained small and midsize construction companies as we have today, or fewer, large regional and national organizations. The choice is for individual construction professionals to make.

Continuing Education versus On-The-Job-Training

If you are gaining experience at work, why should you be interested in continuing education? Long hours are common in our industry and should therefore generate a lot of experience. But if that is the case, everyone is getting the same amount of experience. So, in order to get ahead, one needs to add something else to the on-the-job experience. Moreover, work experience is often similar every day, to the extent that, as mentioned earlier, some will tell you they have 25 years of construction experience when they actually have five years of experience and 20 years of repetition. Experience is not enough to progress your career beyond your peers. You need to add some "new" information to on-the-job learning to advance your professional development. You would not go to a doctor 25 years out of medical school if you knew he or she did not read medical journals and textbooks to keep up on the latest medical developments. The same is true of construction professionals. Working long hours may help you keep up with the competition, but to get ahead of the competition you need to add new information to your experience. The process is referred to as "professional development" and takes the form of a formal professional development plan.

Self-Directed Professional Development

For those who do not have access to company-sponsored education programs but wish to advance in the construction industry, there is an option: a self-directed professional development plan. The process includes personal development because your people skills are as important as your business skills, management skills, and technical skills. The approach recommended is a multiyear activity which you can develop on your own, or preferably with the assistance of a mentor, successful friend, or associate that can help you evaluate your current status and encourage you during the process. The next step is to identify subjects or areas that need improvement or should be learned. Then create a multiyear self-directed plan and pursue it diligently. The authors recommend planning for three years because the objective is not just to learn a few things, but in-depth learning of volumes of information to accelerate your career to new heights. Too many construction professionals argue they are too old to go back to school, but in a self-directed plan, school comes to you.

A formal professional development plan is intended to advance a person well beyond their current level of understanding through accelerated learning over and above the work and life experiences that everyone gains each day. It is up to the individual to decide which areas they wish to advance, both professionally and personally. You should determine the skills and knowledge you wish to enhance, or decide which entirely new skills or knowledge you want to possess.

Establish a Baseline

To establish a baseline, you may benefit by seeking out a test of skills, interests, and/or personality, such as Myers-Briggs, DISC (which stands for the Dominance, Influence, Steadiness, and Compliant personality types), or one of the many other tests available. You can learn a lot about your preferred learning methods, aptitudes, motivations, and interests, some of which you may not have had the opportunity to think much about. For a sample of a technical skills inventory, please refer back to Chapter 16. The objective is to better understand where you are now, in order to design a path to where you wish to be in the future. You need to consider which areas you choose to advance in, develop processes to get you there, and schedule specific learning activities to accomplish these goals. You will need to decide how much time and effort you intend to devote to this plan. A plan like this relies heavily on reading and independent study, and could include selected seminars and possibly short courses. You should set your goals in harmony with your ambition and the time commitment you wish to invest in your future. Being realistic is critical to your plan's success.

Time Commitment and Self-Motivation

You will be investing considerable time and energy to accelerate your growth, so your thoughts and feelings are critical to the initial decisions and should influence the content and "texture" of the plan. A multiyear professional development plan should be flexible because the pace of professional and personal development is set by you, and the learning opportunities and bodies of knowledge change over time. To be effective, this plan needs to be customized to your needs, which may also evolve over time. You should be prepared to alter the plan to accommodate any changes as you progress. The plan should be reviewed annually and updated as needed.

It is essential to determine, at the onset, how many hours per week you intend to devote to this continuing education effort. What gets measured gets done. Deciding to "fit it in" when you can or when you have time is a recipe for failure. If you are not ready to commit and block off a specific number of hours per week to the effort, you may want to reconsider your plan before getting started. "Self-directed" means just that. It is not like "attending" school – it is "schooling" yourself. Self-discipline is crucial because working full-time and setting aside time for study is not easy. It is hard work that can disturb your leisure and social time for the duration of the plan. If you commit to a self-directed multiyear plan, it can have amazing results, but these results are not "free." In terms of time investment, it is similar to going to night school. We know that these types of plans work, but do not want to mislead the reader into thinking the implementation will be easy. In many ways, night school is easier because the schedule and paying tuition stimulate your motivation. Here, in comparison, you will need to be self-motivated.

Seek Collaboration and Encouragement

It is worth looking for a peer with similar ambition to join the effort, exchange ideas, learn from one another, and encourage each other; however, it would take some luck to find someone who happens to be in the same frame of mind as you are. An alternative that would also be helpful to you is to be able to talk with someone about your ongoing education activities as you progress. A relative, friend, or colleague may be willing to participate and may find it of interest. We are not suggesting a tutor or mentor, which would be fine but rarely available, but someone to bounce ideas off and ideally someone to encourage you. One of the issues with studying alone is just that: "alone." It is helpful to have someone interested in what you are doing, to share your thoughts and feelings with. With new videoconferencing technologies such as Zoom, Skype, and Teams, the person does not necessarily need to live locally, which opens up wide choices of individuals to collaborate with. It is also meaningful to be able to discuss your efforts with your family, friends, and associates, many of whom may be more interested in what you are accomplishing than you might think.

Learning Methods

The previous chapter (Chapter 16) provided *a sample company-sponsored professional development plan for a leader-in-the-making.* The sample company-sponsored plan presents and discusses the learning methods that the authors recommend. The same methods also apply to a self-directed plan, and include a skills inventory, substantial reading, a personal resource library, trade publications, self-study including communication skills, active participation in trade associations, seminars and conferences, short courses, and measuring succession plan progress. We recommend reviewing the sample plan in the previous chapter and including these methods in company-sponsored plans as well as self-directed plans. Some of these learning methods can be customized to self-directed plans, as discussed next.

Life-long learning is a commitment to continuous advancement in one's professional life. The foundation of continuing education is reading. Getting started you might want to order one to three books, and then others as you are ready for them. Selecting the books and purchasing them can take time, and you should have at least one book ahead of you for the duration of the development plan, to establish and maintain your reading habit. When you start the last book that you have on hand, order the next one or two so they will be there when you need them. A list of books the authors have used in developing professional development programs is included at the end of this chapter. Most subjects can be learned in disciplined self-study.

Seminars and conferences are subject to your budget and should be selected as they become available, on topics that match your needs. It is recommended that you attend two events per year, if possible. Most are announced months in advance, so it is an ongoing selection process. People you work with are a good source for suggestions, and if your firm has a personnel director or human resources manager, they may be able to help. Short courses may or may not have a place in your plan and may depend on what is available in your area, or you can attend some virtually online. There are costs involved, but coursework can be effective in multiyear plans if you can afford the time and expense.

Seek Assistance

There are areas of the business that you may not need expertise in, but where a basic understanding could give you an advantage. For example, if you are involved in field operations you may have little, if any, exposure to the accounting, human resource, or estimating functions of the business. However, as you advance in your career, you will need at least a basic understanding of those aspects of the organization. These are subjects that you can read about, but to gain practical insights into how they function you can also learn from people working in those areas. After developing a draft of your plan, it may be advantageous to ask senior management if they would be willing to review it and make recommendations. Let them know that you value their input and that you are working on advancing your education and skills to improve your value to the organization. If you have kept you company informed about your formal, self-directed professional development plan you can explain that you would like to gain "working knowledge" of other areas of the business and ask if they would arrange a visit and orientation to those departments, or allow you to observe for a day or so. If that seems inappropriate or does not work out, you can ask someone in the department you are interested in to meet for lunch or after work to explain how that area of the business functions. Or you may have an acquaintance in that profession who works in another firm, perhaps not in construction, who would be willing to enlighten you about the subject in general. The point being, someone probably already knows whatever it is that you wish to learn, so it is a matter of seeking out the experienced person. This is part of the "self-directed" aspect of your plan.

Progress Reports

Measuring your progress and establishing benchmarks is an important element in sustaining motivation. A very brief written progress report is appropriate, if only for the self-satisfaction of seeing the measured results of your work. The report could be as simple as a graph showing which milestones you have achieved, which books you have read, how many hours you already have dedicated to your continuing education, and so on. The progress report does not need to be formal and can include anything that delineates progress toward accomplishing the plan, allowing you to celebrate your progress and success. If you have a study partner, mentor, or collaborator, it might be in order to produce a slightly more detailed report because it helps to "reinforce" the collaboration. A progress report may facilitate and stimulate continuing support and would certainly be of interest to those involved in your plan and to family and friends interested in your success. A multiyear effort of this nature is not confidential; keeping others informed welcomes their participation and emotional support, and in many cases generates admiration and good wishes.

Getting Started: A Sample Self-Directed Professional Development Plan (Figure 17.2)

As mentioned earlier, the cornerstone of professional development is reading; therefore, knowing what to read is important to the busy construction professional. The initial subjects might include management, leadership, business development, corporate development, economic forecasting, personal development and/or business classics. Those interested in advancement should be reading a minimum of six non-fiction books per year as part of their accelerated learning plan.

SAMPLE
3-YEAR SELF-DIRECTED PROFESSIONAL DEVELOPMENT PLAN

Name of Candidate: Year: From ____ To ____

Name of Plan Manager/Mentor:

Introduction
This plan is intended to advance my education, career, and professional capabilities.

Subjects
I will advance my business skills and proficiency in: Communication, Negotiation, Motivation of others, Management and Leadership skills. (You select the subjects).

I will advance my technical skills and proficiency in: Change Order Management, Subcontractor Management, Equipment Management, and Customer Service. (You select the subjects).

I will gain a basic understanding of: Estimating, Pricing, Accounting, Marketing, and Dispute Resolution. (Depending on your position and interests, this list might include: Human Resources, Bonding, Insurance, Risk Management, or any aspect of the construction business).

Time Commitment
I will spend a minimum of eight hours per week working on continuing education. (You select the number of hours and then stick to it. Also specify your intention of evenings, weekends, or both.)

Methodology
The primary learning method will be reading, supplemented with interviews, observations, seminars, and potentially short course(s).

First Year Activities
- I will read a minimum of six books each year of the plan.
- I will develop a personal reference library of books and articles, highlighting information of interest by making notes directly on book pages, articles, and journals, and filing the material for lifetime future reference.
- I will subscribe to and read Businessweek magazine (or HBR, Forbes, The Economist, etc.), Engineering News Record, and will scan/read at least one trade publication from the office weekly.
- I will volunteer at the office to participate in our trade association functions and become familiar with the local construction community.
- I will ask my Uncle _____ (or someone) if he will consider advising me and collaborating on my plan.

1/2

- I will advise senior management of my engagement in a three-year Self-Directed Professional Development Program and seek approval.
- I will develop a progress chart of planned activities and show accomplishments quarterly, sharing it with my uncle and other associates.

Second Year Activities
- Continue the first-year activities.
- I will visit with or interview people knowledgeable in my areas of interest listed above.
- I will attempt to attend one or two industry seminars. I will seek guidance at the office for recommended topics of interest, and hopefully be granted a company sponsorship.
- I will search for a short course appropriate to my work and seek company approval and sponsorship to attend.

Third Year Activities
- Continue first and second year activities.
- Double my efforts to attend at least two industry seminars.
- Double my efforts to interest the company in sponsoring a short course appropriate to my position.
- Plan a celebration of completion and announce widely to my family, friends and associates.
- I will commit to continue "Lifelong Learning" in my personal and professional development after plan completion.

Initial Reading List
To start the learning process and maximize achievements from my initial efforts, I will start with some recommended easy-to-read books that address subjects I intend to advance in.

Personal Development: Body Language; Juluis Fast – A classic
Unlimited Power; Anthony Robbins – For self-motivation

Leadership: Leadership Secrets of Attila The Hun; Wess Roberts – A classic
The 7 Habits of Highly Effective People; Steven Covey – A classic

Management: The One Minute Manager; Kenneth Blanchard – A classic

Business Development: Raving Fans; Kenneth Blanchard – A classic

2/2

Figure 17.2 Three-year self-directed development plan.

Books the Authors Have Used in Developing Professional Development Programs

(Note: The authors tend to favor easy-to-read classics. Some may be out of print. Readers are encouraged to use this list as an initial guide to topics of interest to you, and substitute newer works if preferred. The list covers more topics than one person may be interested in, so select books on topics appropriate to your plan.)

- *Atomic Habits: An Easy & Proven Way to Build Good Habits & Break Bad Ones*; James Clear
- *Attitude is Everything: Change Your Attitude... Change Your Life*; Jeff Keller
- *Balance Sheet Basics: Financial Management for Nonfinancial Managers*; Ronald C. Spurga
- *Be Your Own Coach*; Barbara Braham and Chris Wahl
- *Creating The Accountable Organization*; Mark Samuel
- *Dealing with Difficult People: 24 Lessons for Bringing Out the Best in Everyone*; Dr. Rick Brinkman and Dr. Rick Kirschner
- *Discipline: Training the Mind to Manage Your Life*; Harris kern, Karen Willi
- *Essentialism: The Disciplined Pursuit of Less*; Greg McKeown
- *Excellence Wins: A No-Nonsense Guide to Becoming the Best in a World of Compromise*; Horst Schulze
- *Fierce Conversations: Achieving Success at Work and in life, One Conversation at a Time*; Susan Scott
- *Getting Past No: Negotiating in Difficult Situations*; William Ury
- *Getting to Yes*; Roger Fisher, William Ury
- *Gung Ho! Turn On the People in Any Organization*; Ken Blanchard, Sheldon Bowles
- *How Full is Your Bucket? Positive Strategies for Work and Life*; Tom Rath, Donald O. Clifton
- *How to Read a Financial Report*; John A. Tracy, CPA
- *How to Win Any Argument*; Robert Mayer
- *How to Win Friends & Influence People*; Dale Carnegie
- *Keeping Employees Accountable for Results*; Brian Cole Miller
- *Leadership 101: What Every Leader Needs to Know*; John C. Maxwell
- *Leadership Mastery: How to Challenge Yourself and Others to Greatness*; Dale Carnegie
- *Managing Assertively*; Madelyn Burley-Allen
- *Managing Management Time*; William Oncken
- *Managing Performance to Maximize Results*; Harvard Business School Press
- *Managing Sideways*; Price Pritchett
- *Managing the Profitable Construction Business*; Thomas C Schleifer, Ph.D.
- *Maximum Achievement*; Brian Tracy

- *One Minute for Yourself*; Tim Blanchard, Ph.D. and Spenser Johnson, M.D.
- *Outliers: The Story of Success*; Malcolm Gladwell
- *Patton On Leadership*; Alan Axelrod
- *People Skills: How to Assert Yourself, Listen to Others, and Resolve Conflicts*; Robert Bolton, Ph.D.
- *Principle-Centered Leadership*; Stephen R. Covey
- *QBQ! The Question Behind the Question*; John G. Miller
- *Reading People*; Jo-Ellan Dimitrius, Ph.D., Wendy Patrick Mazzarella
- *Service Excellence*; Price Pritchett
- *Smart Questions: The Essential Strategy for Successful Managers*; Dorothy Leeds
- *Start With No*; Jim Camp
- *Strategic Management in Action*; Mary Coulter
- *Taking Control of Your Time*; Harvard Business School Press
- *Teamwork: The Team Member Handbook*; Price Pritchett
- *The Art of Talking so that People Will Listen*; Paul W. Swets
- *The Definitive Book of Body Language*; Allan and Barbara Pease
- *The Encyclopedia of Business Letters, Faxes, and E-Mails*; Robert W. Bly and Regina Ann Kelly
- *The Leadership Challenge: How to Make Extraordinary Things Happen in Organizations*; James M. Kouzes and Barry Z. Posner
- *The Miniature Guide to Critical Thinking-Concepts and Tools (Thinker's Guide)*; Dr. Richard Paul, Dr. Linda Elder
- *The Only Negotiating Guide You'll Ever Need: 101 Ways to Win Every Time in any Situation*; Peter B. Stark, Jane Flaherty
- *The Power of Positive Confrontation*; Barbara Pachter
- *The Sound of Your Voice*; Dr. Carol Fleming
- *The Unwritten Laws of Business*; W. J. King
- *Too Nice for Your Own Good*; Duke Robinson
- *Transform Your Workplace*; Lynda Ford
- *Unleash the Warrior Within*; Richard J. Machowicz
- *What do I Say Next? Talking Your Way to Business Success*; Susan RoAne
- *What They Don't Teach You at Harvard Business School*; Mark H. McCormack
- *You Can Read Anyone (Never be Fooled, Lied to, or Taken Advantage of Again)*; David, J. Lieberman, Ph.D.
- *Zapp! The Lighting of Empowerment*; William C. Byham

Review Questions. Check All That Apply.

1   An organization that does not regularly renew skills and learn new ones is destined to:

   a   Replicate itself indefinitely

   b   Miss opportunities for improvement

   c   Fall behind efficiency and productivity of more advanced competitors

   d   All of the above

2 How do the authors answer the question on how much does continuing education cost?
   a $1,000 a year per person
   b It depends on the size of the organization
   c As much as an organization can afford
   d Less than you would expect

3 Even modest education and training efforts provide huge returns in the form of:
   a Performance enhancement
   b Employee motivation
   c Productivity improvements
   d All of the above

4 What do the authors consider as the cornerstone of professional development?
   a Reading
   b Mentoring
   c Hands-on practice
   d Short courses

5 How many years do the authors recommend for a self-directed professional development plan?
   a One year
   b Two years
   c Three years
   d Four years

Critical Thinking and Discussion Questions

1 Explain in your own words what you see as the need for continuing education.
2 Outline the advantages of a company-sponsored professional development plans.
3 Explain the development and use of a self-directed professional development plan.
4 Discuss the information covered in the section on "continuing education" versus "on-the-job-training."
5 Expand on what you might do concerning the section on seek assistance.

Answers to the Review Questions

Chapter 1: 1c. 2b&d. 3d. 4c. 5d.
Chapter 2: 1c. 2d. 3d. 4c&d. 5b.
Chapter 3: 1d. 2d. 3b. 4c. 5a.
Chapter 4: 1b. 2a. 3d. 4d. 5d.
Chapter 5: 1d. 2d. 3b. 4a. 5d.
Chapter 6: 1a. 2c. 3c. 4a. 5b.
Chapter 7: 1d. 2b. 3c. 4d. 5c.
Chapter 8: 1a. 2d. 3c. 4c. 5b.
Chapter 9: 1d. 2b. 3c. 4b. 5d.
Chapter 10: 1c. 2b. 3d. 4d. 5c.
Chapter 11: 1a. 2c. 3d. 4d. 5d.
Chapter 12: 1c. 2c. 3b. 4d. 5d.
Chapter 13: 1c. 2b. 3d. 4d. 5d.
Chapter 14: 1c. 2b. 3d. 4a. 5c.
Chapter 15: 1c. 2d. 3a. 4b. 5a.
Chapter 16: 1d. 2c. 3d. 4c. 5d.
Chapter 17: 1a. 2c. 3d. 4a. 5d.

Index

Printed in the United States
by Baker & Taylor Publisher Services

Printed in the United States
by Baker & Taylor Publisher Services